MAGNETOELECTRICITY
IN COMPOSITES

MAGNETOELECTRICITY IN COMPOSITES

Editors

Mirza I. Bichurin
Dwight Viehland

PAN STANFORD PUBLISHING

Published by

Pan Stanford Publishing Pte. Ltd.
Penthouse Level, Suntec Tower 3
8 Temasek Boulevard
Singapore 038988

Email: editorial@panstanford.com
Web: www.panstanford.com

British Library Cataloguing-in-Publication Data
A catalogue record for this book is available from the British Library.

Magnetoelectricity in Composites

ISBN 978-981-4267-79-3 (Hardcover)
ISBN 978-981-4267-83-0 (eBook)

Printed in the USA

Contents

Preface

The concept of magnetoelectricity has been with us for over 60 years. It was first theorized in the 1950s by Landau, Lifshitz, and Dzyaloshinskii, and finally realized in Cr_2O_3 during the late 1950s by Astrov in Russia and Rado in USA. For its first 50 years as a discipline of study, the magnetoelectric (ME) effect remained a curiosity: simply due to the miniscule exchange between magnetic and polar subsystems. However, by the turn of the millennium, various groups were beginning to unlock giant magnetoelectricity in magnetostrictive and piezoelectric composites. It seems that progress in modern materials science and engineering is often inseparably connected to advancements in unique composite materials.

Only in the last few years the ME effect in magnetostriction–piezoelectric composites has achieved its potential for a wide range of practical applications: this hope was advanced by the discovery of strong magnetization–polarization interactions that are orders of magnitude larger than anything previously known. The ME effect in composites is a so-called *product property*, resulting from the action of the magnetostrictive phase on a piezoelectric one that is elastically bonded to it, or vice versa. Appropriate choice of phases with high magnetostriction and piezoelectricity has allowed the achievement of ME voltage coefficients necessary for engineering applications over a wide frequency bandwidth including the electromechanical, magnetoacoustic, and ferromagnetic resonances regimes.

The authors of this book have attempted to set as their goal to bring together numerous contributions to the field of ME composites that have been reported since the beginning of the new millennia. We hope to provide some assimilation of facts into establish knowledge for readers new to the field, so that the potential of the field can be made transparent to new generations of talent to advance the subject matter.

The book is composed of seven chapters.

The structure of the book is as follows. Chapter 1 reviews ME interaction in magnetically ordered materials. Chapter 2 dwells on the effective medium approach to modeling of ME coupling in the low-frequency range. The expressions for ME voltage coefficients are obtained using the material parameters of composite phases: both multilayer and bulk composites are considered. A giant Maxwell–Wagner relaxation is predicted at ultra-low frequencies for a two-phase composite. Section 3 focuses on ME effects in the electromechanical resonance (EMR) region. This chapter presents exact solutions to equations of motion for layered structures. The obtained expressions demonstrate an enhancement of the ME voltage coefficients in the EMR region. Chapter 4 presents the Green function method for calculating the ME coefficients. Some approximations are discussed for bulk composites and two-phase composites of alloys and piezoelectric materials. The size and shape-dependent scaling behavior of the ME effect is explained by giving an example of nanostructured composite thin films. Equivalent-circuit method and ME low-frequency devices are addressed in Chapter 5. Possibilities of application of ME composites for making ac magnetic field sensors, current sensors, transformers, and gyrators are discussed. Chapter 6 focuses on important applications of ME composites in the microwave range. Investigations of a ferromagnetic resonance (FMR) line shift by an applied electric field are discussed for layered ferrite–piezoelectric structures. Layered composites are shown to be of interest for applications as electrically tunable microwave phase-shifters, devices based on the FMR, magnetic-controlled electrooptical and/or piezoelectric devices, and electrically readable magnetic (ME) memories. Chapter 7 is devoted to theory and modeling ME effect in nanostructures. This Chapter shows that the parameters of the structure depend on the mismatch and substrate effects.

The authors are thankful to their colleagues V. Laletin, R. Petrov, Yu. Kiliba, A. Filippov, J.-F. Li, and A. Tatarenko (amongst many more) for important contributions to this work. Thanks to many for their talented contributions and the pleasure that we all have had in advancing a renaissance in the field of magnetoelectricity: maybe let us hope that all of our works will come together in the near future, to achieve the full potential of magnetoelectricity for humankind.

M.I. Bichurin   August 2011
D. Viehland

Chapter 1

Magnetoelectric Interaction in Magnetically Ordered Materials (Review)

M.I. Bichurin

Institute of Electronic and Information Systems, Novgorod State University,
173003 Veliky Novgorod, Russia
Mirza.Bichurin@novsu.ru

This chapter reviews magnetoelectric (ME) interaction in magnetically ordered materials. We discuss the ME properties of magnetostrictive-piezoelectric composites to create new ME composites with enhanced ME couplings that would enable them for application in functional electronics devices. One of the main objectives is a comparative analysis of ME composites that have different connectivity types. It was significant to note that the relative simplicity of manufacturing multilayer composites with a 2–2 type connectivity having giant ME responses is an important benefit. In addition, composites with 3–0 and 0–3 connectivity types are also easy to make using a minimum monitoring of the synthesis process. The ultimate purpose of theoretical estimates is to predict the ME parameters — both susceptibility and voltage coefficients

Magnetoelectricity in Composites
Edited by Mirza I. Bichurin and Dwight Viehland
Copyright © 2012 Pan Stanford Publishing Pte. Ltd.
www.panstanford.com

— as these are the most basic parameters of magnetoelectricity. The ME effects occur over a broad frequency bandwidth, extending from the quasi-static to millimeter ranges. This offers important opportunities in potential device applications.

The magnetoelectric (ME) effect in a material requires that an electric polarization be induced by an applied external magnetic field, or vice versa a magnetization induced by external electric field. Linear state functions that define these cross couplings can be given respectively as

$$P_i = \alpha_{ij} H_j, \tag{1.1}$$

$$M_i = \alpha_{ji}/\mu_0 E_j, \tag{1.2}$$

where P_i is the electric polarization, M_i the magnetization, E_j and H_j the electric and magnetic fields, α_{ij} the ME susceptibility tensor, and μ_0 the permeability of vacuum. The ME effect in solids was first theoretically predicted by Landau and Lifshitz in 1957 [1], calculated by Dzyaloshinskii in Cr_2O_3 [2], and experimentally observed by Astrov [3] and Rado *et al.* [4].

When a material is placed in uniform magnetic and/or electric fields, the change in the Gibbs free energy density can be expressed as [1]

$$dF = -P_i dE_i - \mu_0 M_i dH_i. \tag{1.3}$$

From which it is possible to obtain the following thermodynamic relationships for the dielectric polarization and magnetization:

$$P_i = -(\partial F/\partial E_i)_{H,T}, \tag{1.4}$$

$$\mu_0 M_i = -(\partial F/\partial H_i)_{E,T}, \tag{1.5}$$

where T is the absolute temperature. If we assume that the electric susceptibility (χ^E) and magnetic susceptibility (χ^M) are, respectively, independent of their primary ordering fields E and H, we can obtain the following free energy expression for a linear dielectric and magnetic system that has an ME exchange between the subsystems, given as

$$F = -1/2\, \chi_{ij}^{E} E_i E_j - 1/2\, \chi_{ij}^{M} H_i H_j - \alpha_{ij} E_i H_j. \qquad (1.6)$$

The first term on the right is the electrical energy stored on application of E, the second the magnetic energy stored on application of H, and the third the bilinear coupling between the magnetic and polar subsystems on simultaneous application of E and H. From Eq. 1.6, we can then obtain the following expressions for the total polarization and magnetization induced by simultaneous application of E and H:

$$P_i = \chi_{ij}^{E} E_j + \alpha_{ij} H_j, \qquad (1.7)$$

$$M_i = \chi_{ij}^{M} H_j + \alpha_{ji}/\mu_0 E_j. \qquad (1.8)$$

The ME susceptibility is a second-rank axial tensor, involving an exchange between polar (E_j) and axial (H_j) vectors. It is unique in these regards to the dielectric susceptibility and magnetic permeability, which are both second rank polar tensors. This is an important point, because as a consequence, the values of the components of the ME susceptibility tensor will be dependent on the magnetic point group symmetry, rather than merely on the crystallographic one.

Dzyaloshinskii theoretically showed that among materials of known magnetic point group symmetry that there was at least one crystal — chrome oxide — in which a ME effect [2] should be observed. In 1960, Astrov experimentally verified this previously predicted ME coupling in Cr_2O_3 [3], and reported the values of the longitudinal and transverse ME susceptibility coefficients that are given by Eq. 1.2. These measurements were performed by measuring the ac magnetic moment induced in the sample by application of an ac electric field driven at a frequency of $f = 10$ kHz. Subsequently, Rado *et al.* [4] reported the magnetic field induced polarization of Cr_2O_3 given by Eq. 1.1. Said studies were done using a simple procedure. Electrodes were deposited on both ends of a single crystal bar-shaped sample. The crystals were then located between the poles of an electromagnet, placed in a vacuum chamber, and an electrometer was used to collect the charge induced across the crystal in response to an applied magnetic field. Shortly, thereafter, this same measurement methodology was used by Asher *et al.* [5] to study $Ni_3B_7O_{13}I$.

The structure–property relations of the matrix of ME coefficients has been studied for some systems [4]. Classification of ME materials has been carried out for various magnetic point groups by Schmid, and the symmetry of ME properties determined by application of Neumann's law [9]. It is known that the property matrix is null except for the case of magnetically ordered materials: in which case, the matrix of coefficients is anti-symmetric. Theoretical models based on experimental ME data have been presented for a few material systems [5–8], and interpretation of the ME effect have been attempted using statistical and phenomenological approaches [10, 11]. A Landau-Ginzburg approach has been also applied to study ME effects in ferrimagnets [14], where the magnetic energy density is decomposed into a power series in terms /of the spontaneous magnetization and the spatial derivative thereof. In this case, there is a specific symmetry relationship between the matrix of coefficients in the higher (paramagnetic) and lower (ferromagnetic) temperature phases. Readers should also recognize that important contributions to our understanding of ME materials have been made by theoretical developments of Shavrov [15], Alexander and Shtrikman [16], and Asher [17].

Electromagnetic wave transmission in a ME media has been studied [12, 13]. Brown *et al.* [18] have reported a theoretical value for the upper limit of the ME susceptibility. These authors showed that

$$F + (1/2)\chi_{jj}^d H_i^2 \le 0, \tag{1.9}$$

where χ_{jj}^d is the diamagnetic susceptibility. By taking into consideration of Eq. 1.6, the following inequality results:

$$(1/2)\chi_{ii}^E E_i^2 + \alpha_{ij} E_i H_j + (1/2)\chi_{jj}^p H_j^2 \ge 0, \tag{1.10}$$

where $\chi_{jj}^p = \chi_{jj}^m - \chi_{jj}^d$ is the paramagnetic susceptibility. Taking into account that Eq. 1.10 is a positive-defined inequality, and that $\chi_{ii}^E \ge 0$ and $\chi_{jj}^p \ge 0$, one can arrive at

$$\alpha_{ij} < (\chi_{ii}^E \chi_{jj}^p)^{1/2}. \tag{1.11}$$

If we now suppose that the diamagnetic component is small relative to the paramagnetic one (i.e., we limit ourselves to the case of materials with localized magnet moments), we must consider the case that $\chi_{jj}^{p} \approx \chi_{jj'}^{m}$, and therefore

$$\alpha_{ij} < (\chi_{ii}^{E}\chi_{jj}^{m})^{1/2}. \tag{1.12}$$

An analogous relation can be obtained on the basis of thermodynamic theory [18]

$$\alpha_{ij} < (\varepsilon_{ii}\mu_{jj})^{1/2}, \tag{1.13}$$

where ε and μ are relative dielectric permittivity and magnetic permeability, respectively. Asher and Janner have predicted the upper limits of the ME susceptibility for some point groups [19]. For known material systems, the theoretical upper limit is known to exceed the experimentally measured values.

In addition, Rado [20, 21] discovered a linear ME effect in ferrimagnet $Ga_{2-x}Fe_{x}O_{3}$. Interestingly, in this unusual system, this ME effect was explained by the simultaneous presence of piezoelectric and piezomagnetic properties, where one striction effect was coupled to the other. In ferrum borate $FeBO_{3}$ [22], quadratic ME effects mediated by piezoelectric–piezomagnetic dual couplings has also been reported. These were the first results which suggested that the Dzyaloshinsky-Moriya spin-lattice exchange may not be the only mechanism by which ME effects could be achieved: it was an intellectual precursor to magneto-elasto-electric interactions in composites.

From the 1950s to the present time, many single-phase ME single crystals have been studied: see references [23–29] for a selection of materials properties. Most of these materials have ME effects only at temperature considerably below that of 300 K. This is due to materials having either a low Neel temperature or a low Curie point: there have been no reports of a material with simultaneously large polarizations and magnetizations at or above room temperature. This is important because the ME tensor coefficients are vanquished as soon as the temperature approaches the transition point at which one order parameter transforms to the disordered (para) phase. Furthermore, even at low temperatures, single-phase ME crystals have very small values of the ME coefficients: making them impotent in practical use. Fortunately, composite materials of ferrite and piezoelectric phases exist which have simultaneously large polarizations and magnetizations to temperatures much higher than normal ambient ones: offering to date the only practical approach to ME applications. We now turn our attention in the remainder of this book to ME composites, which are based on magneto-elasto-electric interactions.

1.1 PROPERTIES OF COMPOSITES

Composite materials offer the opportunity to engineer properties that are not available within any of its constituent phases. First, composites have the conventional group of colligative properties which includes, for example, density and stiffness: where composite quantitative adjectives are determined by individual component adjectives, and their volume or weight fractions.

However, composites can also have a (more interesting) second group of properties, which are not intrinsic to its constitutive phases: these are the so-called product tensor properties, first proposed by Van Suchtelen [30, 31]. The appearance of new properties in a composite, which were not present in any individual phase that went into the composite, can be explained as follows. If one of the phases facilitates transformation of an applied independent variable A into an effect B, the inter-relationship of A and B can be characterized by the parameter $X = \partial B/\partial A$ Then, if the second phases converts the variable B to an effect C, the inter-relationship of B and C can be characterized by $Y = \partial C/\partial B$. Then, A must be able to be transformed into C, whose transformation can be characterized by a parameter that is the product of the component parameters: $\partial C/\partial A = (\partial C/\partial B)(\partial B/\partial A) = YX$. Thus, a composite of the two materials must posses a new property which can convert A into C: this property only results from the action of one phase on the other, and is absent in each phase when physically separated from the other.

It is fair to note that B.D.H. Tellegen had already in 1948 proposed a device based on ME composites, which was later named the Tellegen's gyrator [58]. Composites with strong ME effects were not known 60 years ago, and therefore Tellegen's conjecture ME device was not practically realized.

Finally, with regards to any composite materials, there are many possible ways to vary their physical properties, simply by changes and optimizations in the construction and dimensionality of the composite design.

1.2   ME COMPOSITES

The ME effect in composites is as result of elastically coupled piezoelectric and piezomagnetic effects. The mechanism of ME effects in hybrid composites is as follows: the piezomagnetic material is deformed under an applied magnetic field. This deformation results in a mechanical voltage that acts on the piezoelectric component, and hence induces an electric polarization change in the material via piezoelectricity. Obviously, the converse effect is also possible: an applied electric field causes the piezoelectric component to deform, resulting in mechanical voltage acting on the piezomagnetic material, generating a change in magnetization. Either directly or conversely, the net effect is a new tensor property of the composite — the ME effect: which consists of an electric polarization induced by an external magnetic field, and a magnetization induced by an external electric field.

Most of the known magnetically ordered materials have some measurable magnetostrictive effect: as it is a fourth rank polar tensor property, and thus must be exhibited by all crystal classes. However, the piezomagnetic effect is not required by symmetry to be present in all crystal classes [32]; since it is a third rank axial tensor, its property matrix can become null by action of specific symmetry operations. As magnetostriction is the most prevalent form magnetoelastic coupling, deformations (ε_{ij}) induced by external magnetic fields (H) most frequently depend quadratically on field strength, rather than linearly: this is the definition of magnetostriction, i.e., $\varepsilon_{ij} = Q_{ijkl}H_kH_l$, where Q_{ijkl} is the magnetostriction coefficient. This fact makes use of composites difficult in device applications which require linearity over large field ranges. Linearization can only be achieved under applied dc bias magnetic fields. In this case, the ME effect will be close to linear as long as the range of ac magnetic fields remains small in comparisons to that of the superimposed dc magnetic bias.

ME composites were prepared for the first time by van den Boomgard and coworkers. They used unidirectional solidification of a eutectic composition in the quinary system Fe–Co–Ti–Ba–O [33, 34]. This process promotes formation of alternating layers of magnetic spinel and piezoelectric perovskite phases. The unidirectional solidification process requires careful control of composition, in particular when one of the components (oxygen) is a gas. Investigations of said composites have shown that excess TiO_2 (1.5 weight %) allowed obtainment of a ME voltage coefficient of $\alpha_E = dE/dH = 50$ mV/(cmOe): which was considered quite high at that time. However, other compositions showed a much lower ME voltage coefficient in the range of 1–4 mV/(cmOe). In subsequent work, the authors reported a ME voltage coefficient of $\alpha_E = 130$ mV/(come) for the eutectic composition $BaTiO_3$–$CoFe_2O_4$, again prepared by unidirectional solidification [35]. This value was nearly an order of magnitude greater than that for the best single-phase response previously reported which was $\alpha_E = 20$ mV/(cmOe) for single crystal Cr_2O_3. Using the ME voltage coefficient, it is possible to obtain other relevant ME parameters, such as $\alpha = dP/dH = \alpha_E K \varepsilon_0$, where K is the average relative dielectric constant of the composite and ε_0 the permittivity of free space. Using dielectric constants of $K = 500$ for the composite and $K = 11.9$ for Cr_2O_3, we can obtain an estimated value of $\alpha = 7.22 \cdot 10^{-10}$ s/m for the composite, which is approximately in 270 times higher than that of $\alpha = 2.67 \cdot 10^{-12}$ s/m for Cr_2O_3.

In ceramic composites [36], the value for the ME effect of $BaTiO_3$ and $NiFe_2O_4$ alloyed by cobalt and manganese was reduced relative to that prepared by unidirectional solidification. The maximum value of the ME voltage coefficient was 24 mV/(cmOe). The authors reported an unusual polarization behavior in which the field polarity was reversed at a Curie temperature [37]. Using the field created by volume charges in the composite allowed obtainment of make full poled samples. Work [38] focused on ceramic ME composites of the same system, but batched with excess TiO_2, reported the effect of particle size, cooling velocity, and mole concentration of both composite phases. For ceramic composites of $BaTiO_3$ and $Ni(Co,Mn)$ Fe_2O_4, it was found possible to obtain ME voltage coefficients of 80 mV/(cmOe).

There are other early studies of ceramic composites. Bunget and Raetchi observed a ME effect in Ni–Zn ferrite–PZT composites, and studied its dependence of applied magnetic field [39, 40]. The magnitude of the ME voltage coefficient was found to be 3.1 mV/(cmOe). Discussions of the potential for broad-band sensors based on $BaTiO_3$–$NiFe_2O_4$ composites with working frequencies up to 650 kHz should be noted [41]. Composites of Mg–Mn ferrite–$BaTiO_3$ are simultaneously ferroelectric and ferrimagnetic [42]: ferrite/$BaTiO_3$ composites with weight fractions of 30:70, 50:50, 70:30, and 90:10 have all revealed ferroelectric and ferromagnetic hysteresis. Subsequent measurements of the piezoelectric properties of ME composites have shown that the piezoelectric resonance frequency depends on applied magnetic field [43], where the maximum variation of resonance frequency was 0.2% under $H = 875$ kA/m.

Magnetostrictive metals have, more recently, been investigated as alternatives to ferrites in ME laminated composites. Efforts have focused on use of permendur, Terfenol, and Metglas [61–69]. These (Fe, Ni, Co) magnetostrictive alloys offer much larger magnetostrictions. The ME effect in layered composites based on these magnetostrictive alloys and piezoelectric PZT ceramics is significantly larger than those based on ferrites, which is important for engineering applications [61–63]. The maximum ME voltage coefficient reported was about 4 V/(cmOe), which was obtained in multilayer structures of Terfenol and PZT [67–69]. This opens real possibilities for practical devices.

It is also relevant to note that these composites have been considered across a wide frequency bandwidth ranging from near dc to millimeter. Research on microwave ME effects in layered ferrite–piezoelectric composites has been performed [52]. Application of external electric field to the piezoelectric phase is known to induce a shift in the FMR line frequency of the ferrite phase. An analogous effect was reported earlier in bulk ferrite–piezoelectric composites [53]. Detailed analysis of the resonance ME effects in paramagnetic and magnetic ordered materials were performed by Bichurin [54]. Microscopic theories of ME effects were presented in [59, 61] for the magnetic resonance frequency range in magnetically ordered crystals with 3d-electrons. In addition, a theory for low-frequency ME effects and subsequent dispersion with increasing frequency have also been reported [60].

Finally, there are other important works and reviews that should be read. These include the following: (i) a theory of ME effects in homogeneous composites and heterogeneous structures [61]; (ii) applications of ME laminates in devices [55, 56]; and (iii) review articles [73, 74, 75] in which an analysis of the basic operational principles of ME composites is given.

1.3 ESTIMATIONS OF COMPOSITE ME PARAMETERS

Efforts [35] have been devoted to estimation of the ME voltage coefficient for composites, which were based on approximate models. Supposing that the permittivity of barium titanate exceeds considerably that of ferrite, and that the elastic stiffness of both phases are equal, the following relationship has been obtained

$$\alpha_E = (dE/dH)_{\text{composite}} = (dx/dH)_{\text{composite}} \times (dE/dx)_{\text{composite}}$$
$$= {}^{m}v(dx/dH)_{\text{ferrite}} (dE/dx)_{\text{BaTiO}_3},$$

where dx/dH characterizes the material displacement under an applied H, dE/dx is its shape change under an applied E, and ${}^{m}v$ is the ferrite volume fraction.

Using optimistic values for the material parameters of $(dx/x)/dH$ $\approx 6.28 \cdot 10\text{-}9$ m/A, $dE/(dx/x) \approx 2 \cdot 10^9$ V/m, and ${}^{m}v = 0.5$, the maximum ME voltage coefficient was approximated to be $\alpha_E = 5$ V/(cmOe). The more precise expression has also been used in analysis [44]

$$\alpha_E = (dE/dH)_{\text{composite}} = {}^{m}v(dS/dH)_{\text{ferrite}} (1 - {}^{m}v) (dE/dS)_{\text{piezoelectric}}.$$

By taking into account that $dE = dE_3 = g_{33}dT_3$ and $dS = (dT_3)/C_{33}$ (where g_{33} and C_{33} are the piezoelectric and stiffness coefficients of the piezoelectric phase, T the stress, and S the strain), the expression takes the following form:

$$\alpha_E = {}^mv(dS/dH)_{\text{ferrite}} (1 - {}^mv) (g_{33}\ C_{33})_{\text{piezoelectric}}.$$

Estimates for the ME voltage coefficient of $\alpha_E = 0.92$ V/(cmOe) have been obtained using this above formula.

To observe the ME effect in composites, it is necessary to apply a dc magnetic bias in direction of an applied ac magnetic field: this is because the primary magnetoelastic coupling mechanism is magnetostrictive, rather than piezomagnetism. A magnetic bias can be created by an attached permanent magnet, or by a nearby electromagnet. Bunget and Raetchi [39, 40] offered an alternative measurement method. In ME composites, since the electric polarization is a function of both changes in electric and magnetic fields, it is possible to apply them simultaneously and to then measure the polarization. The approach involves measuring the polarization under constant electric field, while varying the magnetic field. Then, the ratio of the change in polarization to the increment in magnetic field yields the ME sensitivity.

Newnham *et al.* [45] offered a classification of composites based on dimensional connectivity types. For example, a composite with one of its phase connected in all three directions (denoted by an index 3) and with a second phase that is isolated having connectivity in no direction (denoted by index 0) was designated as a composite with connectivity type of 3–0. In ME composites containing magnetostrictive ferrites, the ferrite phase has considerably smaller resistance than the piezoelectric one. This leads to a strong dependence of the composite resistance on phase connectivity: the highest resistance occurs for series connection of composite components, whereas the lowest one occurs for parallel. Some ferrites are semiconductors, whose resistance strongly decreases with increasing temperature. To observe ME effects in a composite, it is necessary to permanently pole them under electric field, to get them to demonstrate a piezoelectric effect. However, in ferrites with notable conductivity, it is difficult to get the bulk resistance of the composite sufficiently high to be able to achieve good poling in the piezoelectric phases. However, using a composite with a 0-3 connectivity between ferrite and piezoelectric phases allows for notable increases in the bulk resistivity. Because of this fact, 0-3 composites of ferrite and piezoelectric phases are easier to pole. Consequently, they offer a better practical connectivity form for ME composites made of these materials systems [50, 70].

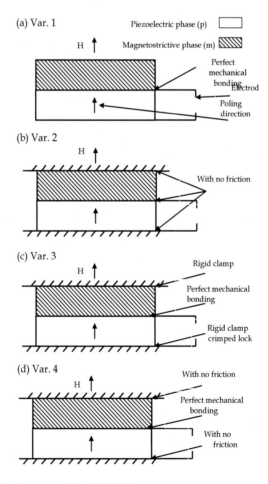

Figure 1.1 Different models of ME bilayer.

Laminate composites, which to date are of most interest for giant magnetoelectricity, have a 2–2 connectivity type: in which case, layers of one phase are stacked on another, in an alternative manner. In this case, each phase has connectivity in two directions in the layer plane, but is not connected with other layer of the same phase which are separated by those of the second phase. This composite type consists of layers that are mechanically connected in series, but which can be electrically connected in either series or parallel as the electrodes can be configured in different manners. Harshe et al. [46, 72] reported calculations of the ME voltage coefficient for composites with a 2–2 connectivity type. They determined that the ME voltage coefficient was the ratio of the electric field induced across the piezoelectric phase to that of the magnetic field applied to magnetostrictive one: i.e., $\alpha_E = {}^pE/{}^mH$.

Figure 1.1 shows various alternative models of ME bilayers with different boundary conditions. The first variant corresponds to a layered composite with ideal mechanical connection between layers. The second bilayer composite variant has boundary conditions representing a clamping of magnetostrictive and piezoelectric layers at both ends, with thin layers of lubricant between all surfaces to minimize friction. The third considers magnetostrictive and piezoelectric layers that are rigidly clamped at both ends. And, the fourth illustrates clamping at both ends, with thin lubricant layers between clamp and sample surfaces to minimize friction. We assume in all four cases, that both phases are perfectly bonded together.

There are some special considerations for the modeling of ME composites that need to be mentioned. First, piezoelectric layers of a composite can be electrically connected in series or parallel. Depending upon which piezoelectric d_{33} or d_{31} and magnetostriction $\lambda_{//}$ or $\lambda_{\perp}$ coefficients that one wishes to take advantage of, different multilayer ME composite configurations can be created [46]. Investigations have been devoted to longitudinal ME effects in bilayer composite based on Terfenol-D [71], in which case a ME voltage coefficient of 1.43 V/cmOe was reported. Second, optimally poled piezoelectric layer effects can be achieved in multilayer structures where ferrite layers are paralleled by intermediate electrodes. In this case, we get a composite with a mechanical connectivity type of 2–2, but with an electrical connectivity type of 3–0. This can be considered as a composite with a mixed connectivity type.

It is often required to predict the effective composite parameters from their constitutive components. Such effective parameters can be determined by the Maxwell-Garnett Eq. 47. For example, in the case of a 0–3 composite, the effective permittivity of a continuous dielectric matrix having a permittivity ε_1 containing isolated second phase particles with ε_2 can be given as follows

$$\varepsilon_{eff} = \varepsilon_1[2\varepsilon_1 + \varepsilon_2 + 2y(\varepsilon_1 - \varepsilon_2)]/[2\varepsilon_1 + \varepsilon_2 - y(\varepsilon_1 - \varepsilon_2)], \quad (1.14)$$

where y is the volume fraction of any additional component.

Calculations of the ME properties of composites have some discrepancy with measured values. Measurement of the ME effect in sintered composites of $NiFe_2O_4$ or $CoFe_2O_4$ and $BaTiO_3$ have been reported [48]. Samples in the form of thin disks were polarized by an electric field applied perpendicular to the sample plane. The ME coefficient was then measured for two cases: (i) transverse fields, where dc and ac magnetic fields were parallel to each other and to the plane of the disk (along directions 1 and 2), which were perpendicular to an ac electric field (along direction 3); and (ii) longitudinal fields, where all three fields (dc and ac magnetic, and ac electric) were parallel to each other and perpendicular to the sample's plane. In general, bulk ceramic composites exhibit values of the ME voltage coefficient that are notably lower than those predicted theoretically by continuum mechanics [49]. One reason for this is the low resistivity of the ferrite phase that was aforementioned, that (i) decreases the electric field which can be applied during poling of the piezoelectric phase, resulting in insufficient polarization; and (ii) leakage current across the electrodes of the composites, which results in an inability of the composite to maintain charge after it has been induced by applied magnetic fields via the piezoelectric effect. The essential advantage of bulk composites is the possibility of reaching the required values of the given parameters using combinations of the constituent phases that have the necessary values of electric and magnetic materials parameters, and alternatively by tuning the effective composite parameters by adjusting the relative phase volume fractions.

Cubic models of ferrite–piezoelectric ME composites with a 3–0 and 0–3 connectivity types have been considered in [49], which allow for numerical computation of ME coefficients. However, this early theoretical model was inadequate: as evidenced by experimental observations that have shown ME voltage coefficients (8×10^{-3} V/(cmOe)) more than two orders of magnitude lower than theoretical predicted one (3.9 V/(cmOe)). Nan *et al.* [62, 63] developed a computational approach to ME effects in bulk composites, based on a Green's functional approach and perturbation theory [62, 63]. The work offered the perspective of a three-phase composite with high mechanical and ME properties.

The ME effect could find wide applications in various types of electronic devices. Possible applications were considered earlier in [24, 25, 29, 51, 57]: microwave phase shifters were considered in [55], ME magnetic field and microwave power sensors in [56]. For such applications, ME composites are necessary (rather than single-phase crystals) due to their high values of ME coefficients and higher working temperature range. However, application of ME composites is hindered by bad reproducibility of effective composite parameters. For example, good mechanical connection across layer interface between phases is necessary to achieve good ME coupling. Furthermore, the constituent phases of a composite should not react with each other. This is a concern in particular for sintered composites based on ferrite–piezoelectric ceramics, as very high temperatures are used in the densification process: avoidance of chemical reactions at interfacial areas can then complicate processing of sintered composites.

1.4 CONCLUSIONS

In this chapter, we discussed the ME properties of ferrite–piezoelectric composites, to create new ME composites with enhanced ME couplings that would enable them for application in functional electronics devices. To address this important scientific and technical goal, a generalization of various theoretical and experimental studies of ME composites has been given. One of the main tasks according to the formulated approach is a comparative analysis of ME composites that have different connectivity types. The relative simplicity of manufacturing multilayer composites with a 2–2 type connectivity having giant ME responses is an important benefit. In addition, composites with 3–0 and 0–3 connectivity types can also be made in considerable quantity by a minimum monitoring of the synthesis process.

Any material with a high piezoelectric constant is a good choice for the piezoelectric phase in ME composites. The most suitable ones are $Pb(Zr,Ti)O_3$, $BaTiO_3$, and $Pb(Mg_{1/3}Nb_{2/3})O3-PbTiO_3$, which is due to their large piezoelectric coefficients. Analogously, any material with a high piezomagnetic coefficient at low magnetic biases is a good candidate for the magnetostrictive phase. The most suitable choices are ferrites and ferromagnetic metals (such as Terfenol-D and Metglas).

Availability of theoretical models for composite properties is necessary to interpret experimental data and to restrict oneself among the multitude of composite configurations. Theoretical estimations of the ME voltage coefficient for series and parallel composite models, and also a cubic model for composites with a 3–0 connectivity type [46, 49], are known. However, as already mentioned, ME voltage coefficient was computed as the ratio of an electric field induced in the piezoelectric phase to the magnetic field applied to the magnetic one: i.e., $\alpha_E = {}^pE/{}^mH$. But, in reality, the internal fields in composite components can be significantly different from the external fields. In particular, formulas [46, 49] predicted a maximum ME voltage coefficient in a pure piezomagnetic phase (i.e., ${}^pv = 0$): which distinctly mismatches reality. In addition, generalized models for composite based on an effective medium method were presented. These offer determination of the effective composite parameters with phase connectivity types of 2–2, 3–0, and 0–3 that are based on exact solutions. The ultimate purpose of any theoretical work must be to predict the ME parameters — both susceptibility and voltage coefficients — as these are the most basic parameters of magnetoelectricity.

It is important to realize that ME effects occur over a broad frequency bandwidth, extending from the quasi-static to millimeter ranges. This offers important opportunities in potential device applications. It makes possible new concepts in sensing, gyrators, microwave communications, phase shifters, just to name a few. It also complicates the understanding of magnetoelectricity, as there are significant changes in its spectra with frequency. There are strong enhancements in the ME coefficients near both the electromechanical and magnetomechanical resonances. In addition, there is the important problem of studying dispersion in the ME parameters over a broad frequency range of $10^{-3} < f < 10^{10}$ Hz. Relaxation parameters depend on connectivity type, composite geometry and structure, and volume fraction of constituent phases.

This book surveys the ME effect in ME composites over a wide frequency range, offering suggestions for making new ME materials with sufficient exchange to enable practical applications. Generalized theoretical and experimental studies will be presented that try to gain advantage by comparing existing solutions with existing data.

Acknowledgments

This work was supported by the Russian Foundation for Basic Research and Program of Russian Ministry of Education and Science.

References

1. L.D. Landau, E.M. Lifshitz, *Statistical Physics*, 3rd ed., Pergamon, Oxford, 1980, 562p.

2. I.E. Dzyaloshinskii, "On the magneto-electrical effect in antiferromagnets," Sov. Phys. JETP, **10**, 628 (1960).

3. D.N. Astrov, "Magnetoelectric effect in chromium oxide," Sov. Phys. JETP, **13**, 729 (1961).

4. V.J. Folen, G.T. Rado, E.W. Stalder, "Anysotropy of the magnetoelectric effect in Cr2O3," Phys. Rev. Lett., **6**, 607 (1961).

5. E. Asher, "The interaction between magnetization and polarization: phenomenological symmetry consideration," J. Phys. Soc. Jpn., **28**, 7 (1969).

6. R.P. Santoro, R.E. Newnham, "Survey of magnetoelectric materials," Technical Report AFML TR-66–327, Air Force Materials Lab., Ohio, 1966.

7. H. Yatom, R. Englman, "Theoretical methods in magnetoelectric effect," Phys. Rev., **B188**, 793 (1969).

8. R. Englman, H. Yatom, "Low temperature theories of magnetoelectric effect," Proceedings of Symposium on Magnetoelectric Interaction in Crystals, USA, 1973, Eds. A. Freeman, H. Schmid, Gordon and Breach Sci. Publ., New York, **17** (1975).

9. H. Schmid, "On a magnetoelectric classification of materials," Proceedings of Symposium on Magnetoelectric Interaction in Crystals, USA, 1973, Eds. A. Freeman, H. Schmid, Gordon and Breach Sci. Publ., New York, **121** (1975).

10. G.T. Rado, "Statistical theory of magnetoelectric effect in an antiferromagnetics," Phys. Rev., **128**, 2546 (1962).

11. W. Opechovski, "Magnetoelectric symmetry," Proceedings of Symposium on Magnetoelectric Interaction in Crystals, USA, 1973, Eds. Freeman A and Schmid H., Gordon and Breach Sci. Publ., New York, **47** (1975).

12. T.H. O'Dell, *The Electrodynamics of Magnetoelectric Media*, North-Holland Publ. Company, Amsterdam, 1970, 304p.

13. R. Fuchs, "Wave propagation in a magnetoelectric medium," Phil. Mag., **11**, 647 (1965).

14. G. Aubert, "A novel approach of the magnetoelectric effect in antiferromagnets," J. Appl. Phys., **53**, 8125 (1982).

15. V.G. Shavrov, "The magnetoelectric effect," Sov. Phys. JETP, **21**, 948 (1965).

16. S. Alexander, S. Shtrikman, "On the origin of axial magnetoelectric effect of Cr_2O_3," Solid State Commun., **4**, 115 (1966).

17. E. Asher, "The interaction between magnetization and polarization: phenomenological symmetry consideration," J. Phys. Soc. Jpn., **28**, 7 (1969).

18. Jr. W.F. Brown *et al.* "Upper bound on the magnetoelectric susceptibility," Phys. Rev., **168**, 574 (1968).

19. E. Asher, A.G.M. Janner, "Upper bounds on the magnetoelectric susceptibility," Phys. Lett., **A29**, 295 (1969).

20. G.T. Rado, "Observation and possible mechanisms of magnetoelectric effect in ferromagnet," Phys. Rev. Lett., **13**, 335 (1964).

21. G.T. Rado, "Present status of the theory of magnetoelectric effects," Proceedings of Symposium on Magnetoelectric Interaction in Crystals, USA, 1973, Eds. A. Freeman, H. Schmid, Gordon and Breach Sci. Publ., New York, **32** (1975).

22. M.I. Bichurin, V.M. Petrov, "Electric field influence on antiferromagnetic resonance spectrum in iron borate," Phys. Solid State, **29**, 2509 (1987).

23. *Magnetoelectric Interaction Phenomena in Crystals*, Eds. A.J. Freeman, H. Schmid, N.-J. Paris, Gordon and Breach, London, 1975, 228p.

24. G.A. Smolenskii, I.E. Chupis, "Magnetoelectrics," Adv. Phys. Sci., **137**, 415 (1982) (in Russian).

25. *Magnetoelectric Substances*, Eds. Yu.N. Venevtsev, V.N. Lyubimov, Nauka, Moscow, 1990, 184p (in Russian).

26. Proceedings of the 2nd International Conference on Magnetoelectric Interaction Phenomena in Crystals (MEIPIC-2, Ascona), Eds. H. Schmid, A. Janner, H. Grimmer, J.-P. Rivera, Z.-G. Ye, Ferroelectrics, **161−162**, 1993, 748p.

27. Proceedings of the 3rd International Conference on Magnetoelectric Interaction Phenomena in Crystals (MEIPIC-3, Novgorod), Ed. M.I. Bichurin, Ferroelectrics, **204**, 1997, 356p.

28. Proceedings of the Fourth Conference on Magnetoelectric Interaction Phenomena in Crystals (MEIPIC-4, Veliky Novgorod), Ed. M.I. Bichurin, Ferroelectrics, **279−280**, 2002, 386p.

29. Proceedings of the Fifth Conference on Magnetoelectric Interaction Phenomena in Crystals (MEIPIC-5, Sudak), Eds. M.Fiebig, V.V. Eremenko, I.E. Chupis, Kluwer Academic Publishers, Netherlands, NATO Sciences Series, 2004, 334p.

30. J. Van Suchtelen, "Product properties: a new application of composite materials," Philips Res. Rep., **27**, 28 (1972).

31. J. Van Suchtelen, "Non structural application of composite materials," Ann. Chim. Fr., **5**, 139 (1980).

32. I.E. Dzyaloshinskii, "The problem of piezomagnetism," Sov. Phys. JETP, **6**, 621 (1958).

33. J. Van den Boomgard *et al.*, "An in situ grown eutectic magnetoelectric composite materials: part I," J. Mater. Sci., **9**, 1705 (1974).

34. A.M.J.G. Van Run *et al.*, "An in situ grown eutectic magnetoelectric composite materials: part II," J. Mater. Sci., **9**, 1710 (1974).

35. J. Van den Boomgard, A.M.J.G. Van Run, J. Van Suchtelen, "Magnetoelectricity in piezoelectric-magnetostrictive composites," Ferroelectrics, **10**, 295 (1976).

36. J. Van den Boomgard, A.M.J.G. Van Run, J. Van Suchtelen, "Piezoelectric-piezomagnetic composites with magnetoelectric effect," Ferroelectrics, **14**, 727 (1976).

37. J. Van den Boomgard, A.M.J.G. Van Run, "Poling of a ferroelectric medium by means of a built-in space charge field with special reference to sintered magnetoelectric composites," Solid State Commun., **19**, 405 (1976).

38. J. Van den Boomgard, R.A.J. Born, "Sintered magnetoelectric composite material BaTiO3Ni(Co, Mn)Fe2O4," J. Mater. Sci., **13**, 1538 (1978).

39. I. Bunget, V. Raetchi, "Magnetoelectric effect in the heterogeneous system NiZn ferrite — PZT ceramic," Phys. Stat. Sol., **63**, 55 (1981).

40. I. Bunget, V. Raetchi, "Dynamic magnetoelectric effect in the composite system of NiZn ferrite and PZT ceramics," Rev. Roum. Phys., **27**, 401 (1982).

41. Bracke L.P.M., R.G. Van Vliet, "Broadband magneto-electric transducer using a composite material," Int. J. Electron., **51**, 255 (1981).

42. R. Rottenbacher, H.J. Oel, G. Tomandel, "Ferroelectrics ferromagnetics," Ceramics Int., **106**, 106 (1990).

43. A.E. Gelyasin, V.M. Laletin, "Magnetic bias influence on resonance frequency composite ceramics ferrite-piezoelectric," Tech. Phys. Lett., **14**, 1746 (1988).

44. A.S. Zubkov, "Magnetoelectric effect in ferromagnetic composites based on piezoelectric and magnetostrictive materials," Elektrichestvo, **10**, 77 (1978) (in Russian).

45. R.E. Newnham, D.P. Skinner, L.E. Cross, "Connectivity and piezo-electric-pyroelectric composites," Mater. Res. Bull., **13**, 525 (1978).

46. G. Harshe, J.O. Dougherty, R.E. Newnham, "Theoretical modelling of multilayer magnetoelectric composites," Int. J. Appl. Electromagn. Mater., **4**, 145 (1993).

47. J.V. Mantese *et al.* "Applicability of effective medium theory to ferroelectric/ferromagnetic composites with composition and frequency-dependent complex permittivities and permeabilities," J. Appl. Phys., **79**, 1655 (1996).

48. J. Van den Boomgaard, D.R. Terrell, R.A.J. Born, H.F.J.I. Giller "An in situ grown eutectic magnetoelectric composite material," J. Mater. Sci., **9**, 1705 (1974).

49. G. Harshe, J.P. Dougherty, R.E. Newnham, "Theoretical modelling of 3-0, 0-3 magnetoelectric composites," Int. J. Appl. Electromagn. Mater., **4**, 161 (1993).

50. J. Zhai, N. Cai, L. Liu, H. Yuan, C.-W. Nan, "Dielectric behavior and magnetoelectric properties of lead zirconate titanate/Co-ferrite particulate composites," Mater. Sci. Eng., **B99**, 329 (2003).

51. M.I. Bichurin, V.M. Petrov *et al.*, "Magnetoelectric materials: technology features and application perspectives," in *Magnetoelectric Substances*, Nauka, Russia 118 (1990) (in Russian).

52. M.I. Bichurin, V.M. Petrov, "Magnetic resonance in layered ferrite-ferroelectric structures," Sov. Phys. JETP, **58**, 2277 (1988).

53. M.I. Bichurin, O.S. Didkovskaya, V.M. Petrov, S.E. Sofronev, "Resonant magnetoelectric effect in composite materials," Izv. vuzov., ser. Physic. **1**, 121 (1985) (in Russian).

54. M.I. Bichurin, "Magnetoelectric resonance effects in paramagnetic and magnetically ordered media at ultrahigh frequencies: doctor's dissertation," Novgorod Politehn. Institute, Novgorod, 1988, 288p (in Russian).

55. R.V. Petrov, "Magnetoelectric microwave phase-shifter: candidate's dissertation," Yaroslav-the-Wise Novgorod State University, V. Novgorod, 1997, 120p (in Russian).

56. Yu.V. Kiliba, "Development of magnetoelectric sensors magnetic field and microwave power based on composites: candidate's dissertation," Yaroslav-the-Wise Novgorod State University, V. Novgorod, 2003, 124p (in Russian).

57. M.I. Bichurin, V.M. Petrov, "Magnetoelectric materials in the microwave range," Yaroslav-the-Wise Novgorod State University, Novgorod, 1998, 154p (in Russian).

58. B.D.H. Tellegen, "The gyrator, a new electric network element," Philips Res. Rep., **3**, 81 (1948).

59. I.S. Nikiforov, "Resonant magnetoelectric effect in chromium oxide and iron borate: candidate's dissertation," Yaroslav-the-Wise Novgorod State University, V. Novgorod, 2004, 166p (in Russian).

60. V.M. Petrov, "Magnetoelectric properties of the ferrite-piezoelectric composites," doctor's dissertation, Yaroslav-the-Wise Novgorod State University, V. Novgorod, 2004, 186p (in Russian).

61. D.A. Filippov, "Magnetoelectric effect in magnetic-ordered crystals with 3d-ion and ferrite-piezoelectric composites in the magnetic and electromechanical resonance range," doctor's dissertation, Yaroslav-the-Wise Novgorod State University, V. Novgorod, 2004, 196p (in Russian).

62. C.-W. Nan, Y. Lin, J.H. Huang, "Magnetoelectricity of multiferroic composites," Ferroelectrics, **280**, 153 (2002).

63. N. Cai, J. Zhai, C.-W. Nan, Y. Lin, Z. Shi, "Dielectric, ferroelectric, magnetic and magnetoelectric properties of multiferroic laminated composites," Phys. Rev., **B68**, 224103 (2003).

64. J.F. Li, D. Viehland, "Giant magneto-electric effect in laminate composites," IEEE Trans. Ultrason. Ferroelectr. Freq. Control, **50**, 1236 (2003).

65. S. Dong, J.F. Li, D. Viehland, "Longitudinal and transverse magnetoelectric voltage coefficients of magnetostrictive/piezoelectric laminate composite: theory," IEEE Trans. Ultrason. Ferroelectr. Freq. Control, **50**, 1253 (2003).

66. S. Dong, D. Viehland, "Giant magneto-electric effect in laminate composites," Phil. Mag. Lett., **83**, 769 (2003).

67. S. Dong, J.F. Li, D. Viehland, "Characterization of magnetoelectric laminate composites operated in longitudinal-transverse and transverse-transverse modes," J. Appl. Phys., **95**, 2625 (2004).

68. J. Ryu, A.V. Carazo, K. Uchino, H.E. Kim, "Magnetoelectric properties in piezoelectric and magnetostrictive laminate composites," Jpn. J. Appl. Phys., **40**, 4948 (2001).

69. J. Ryu, S. Priya, A.V. Carazo, K. Uchino, H.E. Kim, "Effect of magnetostrictive layer on magnetoelectric properties in lead zirconate titanate/terfenol-d laminate composites," J. Am. Ceram. Soc., **84**, 2905 (2001).

70. J. Ryu, A.V. Carazo, K. Uchino, H.E. Kim, "Piezoelectric and magnetoelectric properties of lead zirconate titanate/ni-ferrite particulate composites," J. Electroceram., **7**, 17 (2001).

71. K. Mori, M. Wuttig, "Magnetoelectric coupling in terfenol-d/polyvinylidenedifluoride composites," Appl. Phys. Lett., **81**, 100 (2002).

72. C.-W. Nan, "Magnetoelectric effects in composites of piezoelectric and piezomagnetic phases," Phys. Rev., **B50**, 6082 (1994).

73. M. Fiebig, "Revival of the magnetoelectric effect," J. Phys. D: Appl. Phys., **38**, R1 (2005).

74. M.I. Bichurin, D. Viehland, and G. Srinivasan, "Magnetoelectric interactions in ferromagnetic — piezoelectric layered structures: phenomena and devices," J. Electroceram., **19**, 243 (2007).

75. C.-W. Nan, M.I. Bichurin, S. Dong, D. Viehland, G. Srinivasan, "Multiferroic magnetoelectric composites: historical perspectives, status, and future directions" J. Appl. Phys., **103**, 031101 (2008).

Chapter 2

Effective Medium Approach: Low-Frequency Range

M.I. Bichurin and V.M. Petrov

Institute of Electronic and Information Systems, Novgorod State University,
173003 Veliky Novgorod, Russia
Mirza.Bichurin@novsu.ru

The subject of this chapter is the theoretical modeling of low-frequency magnetoelectric (ME) effect in layered and bulk composites based on magnetostrictive and piezoelectric materials. Our analysis rests on the effective medium theory. The expressions for effective parameters including ME susceptibilities and ME voltage coefficients as the functions of interface parameter, material parameters, and volume fractions of components are obtained. Longitudinal, transverse, and in-plane cases are considered. The use of the offered model has allowed to present the ME effect in ferrite cobalt–barium titanate, ferrite cobalt–PZT, ferrite nickel–PZT, lanthanum strontium manganite–PZT composites adequately for the first time. For layered ferrite–piezoelectric composite the Maxwell-Wagner relaxation of ME susceptibility and ME voltage coefficient, which has Debay character and for ME susceptibility is normal and for ME voltage coefficient is inverse, is found. In ferrite–piezoelectric bulk composite the Maxwell-Wagner relaxation of ME susceptibility and ME voltage coefficient, which for ME voltage coefficient is inverse and for ME susceptibility can be both normal and inverse, is found out.

Magnetoelectricity in Composites
Edited by Mirza I. Bichurin and Dwight Viehland
Copyright © 2012 Pan Stanford Publishing Pte. Ltd.
www.panstanford.com

2.1 MULTILAYER COMPOSITES

A series of difficulties peculiar to bulk composites can be overcome in layered structures. Giant magnetoelectric (ME) effects in layered composites are due to (i) high piezoelectric and piezomagnetic coefficients in individual layers, (ii) effective stress transfer between layers, (iii) ease of poling and subsequent achievement of a full piezoelectric effect, and (iv) ability to hold charge due to suppression of leakage currents across composites with a 2–2 connectivity.

Prior theoretical models based on mechanics and constitutive relationships by Harshe *et al.* [6] were restricted to account for ME voltage coefficients in laminates having ideal mechanical connection at the interfaces between layers. Principal disadvantages of this earlier approach [6] were as follows: (i) For the case of longitudinally oriented fields, the effect of the magnetic permeability of the ferrite phase was ignored. Diminution of interior (local) magnetic fields results in a weakening of ME interactions via demagnetization fields. (ii) The case of fields applied in cross orientations to the ME layer connectivity was not considered, which later experimental investigations revealed large ME responses. And, (iii) imperfection of the mechanical connection between constituent phases was not considered.

In this chapter, we present a summary of a more recent theory of ME laminate composites, which are free from the disadvantages mentioned just above. The approach is based on continuum mechanics, and considers the composite as a homogeneous medium having piezoelectric and magnetostrictive subsystems.

2.1.1 Model and Basic Equations

To derive the effective material parameters of composites, an averaging method consisting of two steps [8–11] should be used. In the first step, the composite is considered as a structure whose magnetostrictive and piezoelectric phases are distinct and separable. ME composites are characterized by the presence of magnetic and electric subsystems interacting with each other.

The constitutive equation for the piezoelectric effect can be given in the following form:

$$P = d \cdot T, \tag{2.1}$$

where P is the electric polarization (a 3×1 matrix), d the piezoelectric modulus (a 3×6 matrix), and T the mechanical stress (a 6×1 matrix). The electric displacement D is then

$$D = d \cdot T + \varepsilon^T \cdot E, \tag{2.2}$$

where E is the electric field (a 3×1 matrix), and ε^T and the stress-free (i.e., $T = 0$) dielectric permittivity of the composite. Under open circuit conditions (i.e., $D = 0$), it follows from Eq. 2.2 that

$$E = (-d/\varepsilon^T)T. \tag{2.3}$$

The constitutive equation for the magnetostriction strain of a material in an applied magnetic field, neglecting hysteresis, can be given as follows:

$$S = b \cdot M^2, \tag{2.4}$$

where S is the strain (a 6×1 matrix), b the magnetoelastic constant (a 6×6 matrix), and M the magnetization (a 3×1 matrix). Under a superimposed dc magnetic bias, the magnetostriction can be considered as a pseudo-linear piezomagnetic effect. If the amplitude of a variable magnetic field is small relative to that of the bias, then the magnitude of the magnetostriction coefficient $(\delta S / \delta H)$ can be accepted as being constant. Then, the pseudo-linear piezomagnetic strain can be expressed as

$$S = q \cdot H, \tag{2.5}$$

where q is the effective linear piezomagnetic coefficient (a 6×3 matrix).

The total strain is then a linear combination of the strains of the individual phases, given as

$$S = s \cdot T + d \cdot E + q \cdot H. \tag{2.6}$$

From Eq. 2.6, the strain and electric displacement tensors of the piezoelectric phase are respectively

$$^{p}S_{i} = {}^{p}s_{ij}{}^{p}T_{j} + {}^{p}d_{ki}{}^{p}E_{k},$$ (2.7)

$$^{p}D_{k} = {}^{p}d_{ki}{}^{p}T_{i} + {}^{p}\varepsilon_{kn}{}^{p}E_{n},$$ (2.8)

where $^{p}S_{i}$ is a strain tensor component of the piezoelectric phase; $^{p}E_{k}$ is a vector component of the electric field; $^{p}D_{k}$ is a vector component of the electric displacement; $^{p}T_{i}$ is a stress tensor component of the piezoelectric phase; $^{p}s_{ij}$ is a compliance coefficient of the piezoelectric phase; $^{p}d_{ki}$ is a piezoelectric coefficient of the piezoelectric phase; and $^{p}\varepsilon_{kn}$ is a permittivity matrix of the piezoelectric phase.

Analogously, the strain and magnetic induction tensors of the magnetostrictive phase are respectively

$$^{m}S_{i} = {}^{m}s_{ij}{}^{m}T_{j} + {}^{m}q_{ki}{}^{m}H_{k},$$ (2.9)

$$^{m}B_{k} = {}^{m}q_{ki}{}^{m}T_{i} + {}^{m}\mu_{kn}{}^{m}H_{n},$$ (2.10)

where $^{m}S_{i}$ is a strain tensor component of the magnetostrictive phase; $^{m}T_{j}$ is a stress tensor component of the magnetostrictive phase; $^{m}s_{ij}$ is a compliance coefficient of the magnetostrictive phase; $^{m}H_{k}$ is a vector component of magnetic field; $^{m}B_{k}$ is a vector component of magnetic induction; $^{m}q_{ki}$ is a piezomagnetic coefficient; and $^{m}\mu_{kn}$ is a permeability matrix.

Prior models assumed that the connection at interfaces between layers was ideal. However, in this chapter, we assume that there is a coupling parameter $k = ({}^{p}S_{i} - {}^{p}S_{i0})/({}^{m}S_{i} - {}^{p}S_{i0})$ ($i = 1,2$), where $^{p}S_{i0}$ is a strain tensor component assuming no friction between layers [11]. This interphase–interface elastic–elastic coupling parameter depends on interface quality, and is a measure of a differential deformation between piezoelectric and magnetostrictive layers. The coupling parameter is $k = 1$ for the case of an ideal interface, and *is* $k = 0$ for the case of no friction.

In the second step of the averaging method to derive the effective materials parameters, the bilayer composite is considered as a homogeneous solid, whose behavior can be described by the following coupled sets of linear algebraic equations:

$$S_{i} = s_{ij}T_{j} + d_{ki}E_{k} + q_{ki}H_{k},$$ (2.11)

$$D_{k} = d_{ki}T_{i} + \varepsilon_{kn}E_{n} + a_{kn}H_{n},$$ (2.12)

$$B_k = q_{ki}T_i + a_{kn}E_n + \mu_{kn}H_n, \tag{2.13}$$

where S_i is a strain tensor component; T_j is a stress tensor component;

E_k is a vector component of the electric field; D_k is a vector component of the electric displacement; H_k is a vector component of the magnetic field; B_k is a vector component of the magnetic induction; s_{ij} is an effective compliance coefficient; d_{ki} is a piezoelectric coefficient; q_{ki} is a piezomagnetic coefficient; ε_{kn} is an effective permittivity; μ_{kn} is a permeability coefficient; and a_{kn} is a ME coefficient.

The simultaneous solution of the coupled sets of linear algebraic equations given in Eqs. 2.7–2.13 allows one to find the effective parameters of a composite.

2.1.2 ME Effect in Free Samples

2.1.2.1 Longitudinal ME effect

Let us consider that the layers of a composite are oriented along the planes (X_1, X_2), and that the direction X_3 is perpendicular to the same plane. In this case, the direction of polarization in a sample coincides with the X_3 axis. If we by choice apply a constant magnetic bias and/or variable magnetic field along the same direction coincidental with that of the polarization,

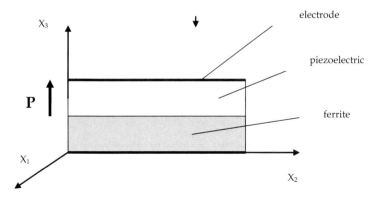

Figure 2.1 Schematic of the bilayer composite structure.

Table 2.1 Non-zero coefficients of piezoelectric and magnetostrictive phases and homogeneous material for longitudinal field orientation.

Piezoelectric phase	
Piezoelectric coefficients	Compliance coefficients
$^{p}d_{15} = {}^{p}d_{24}$	$^{p}s_{11} = {}^{p}s_{22}$
$^{p}d_{31} = {}^{p}d_{32}$	$^{p}s_{12} = {}^{p}s_{21}$
$^{p}d_{33}$	$^{p}s_{13} = {}^{p}s_{23} = {}^{p}s_{31} = {}^{p}s_{32}$
	$^{p}s_{33}$
	$^{p}s_{44} = {}^{p}s_{55}$
	$^{p}s_{66} = 2\left({}^{p}s_{11} + {}^{p}s_{12}\right)$

Magnetostrictive phase	
Piezomagnetic coefficients	Compliance coefficients
$^{m}q_{15} = {}^{m}q_{24}$	$^{m}s_{11} = {}^{m}s_{22} = {}^{m}s_{33}$
$^{m}q_{31} = {}^{m}q_{32}$	$^{m}s_{12} = {}^{m}s_{21} = {}^{m}s_{13} = {}^{m}s_{23} = {}^{m}s_{31} = {}^{m}s_{32}$
$^{m}q_{33}$	$^{m}s_{44} = {}^{m}s_{55} = {}^{m}s_{66}$

Homogeneous material		
Piezoelectric coefficients	Piezomagnetic coefficients	Compliance coefficients
$D_{15} = d_{24}$	$q_{15} = q_{24}$	$^{p}s_{11} = {}^{p}s_{22}$
$d_{31} = d_{32}$	$q_{31} = q_{32}$	$^{p}s_{12} = {}^{p}s_{21}$
d_{33}	q_{33}	$^{p}s_{13} = {}^{p}s_{23} = {}^{p}s_{31} = {}^{p}s_{32}$
		$^{p}s_{33}$
		$^{p}s_{44} = {}^{p}s_{66}$
		$^{p}s_{66} = 2\left({}^{p}s_{11} + {}^{p}s_{12}\right)$

then any resultant electric field will also be parallel to the X_3 axis, as shown in Fig. 2.1. The non-zero components of $^{p}s_{ij}$, $^{p}d_{ki}$, $^{m}s_{ij}$, $^{m}q_{ki}$, s_{ij}, d_{ki}, q_{ki}, α_{kn} for this configuration are summarized in Table 2.1. This summarization supposes that the symmetry of the piezoelectric phase is ∞ m, and that of the magnetic phase is cubic. The following boundary conditions can then be used to solve Eqs. 2.7–2.13.

$$^{p}S_{i} = k\,{}^{m}S_{i} + (1 - k)\,{}^{p}S_{i0}; \; (i = 1,2)$$

$$^{p}T_{i} = -{}^{m}T_{i}(1 - v)/\,v; \; (i = 1,2) \tag{2.14}$$

$$^pT_3 = {}^mT_3 = T_3;$$

$$S_3 = [{}^pS_3 + {}^mS_3(1 - v)]/v,$$

where $v = {}^pv/({}^pv + {}^mv)$ and ^pv and ^mv denote the Poisson's ratio of the piezoelectric and magnetostrictive phases, respectively; and $^pS_{10}$ and $^pS_{20}$ are the strain tensor components for $k = 0$.

To find the effective piezoelectric and piezomagnetic coefficients, it is necessary to consider the composite under both applied electric field $E_3 = V/t$ (V is the applied voltage and t is the thickness of the composite) and applied magnetic field H_3. The effective E in the piezoelectric and H in the magnetostrictive phases are given by: $^pE_3 = (E_3 - (1 - v)^pD_3/\varepsilon_0)/v$, $^mH_3 = (H_3 - v^mB_3/\mu_0)/(1 - v)$. Using continuity conditions for magnetic and electric fields, and using open and closed circuit conditions, one can then obtain the following expressions for the effective permittivity, permeability, ME susceptibility, and longitudinal ME voltage coefficient.

$$\varepsilon_{33} = \{2\,({}^pd_{31})^2\,(v - 1) + {}^p\varepsilon_{33}[({}^ps_{11} + {}^ps_{12})(1 - v)$$
$$+ kv({}^ms_{11} + {}^ms_{12})]\}/\{v[({}^ps_{11} + {}^ps_{12})(1 - v) + kv({}^ms_{11} + {}^ms_{12})]\}, \quad (2.15)$$

$$\mu_{33} = \mu_0\{{}^m\mu_{33}[kv({}^ms_{11} + {}^ms_{12}) + (1 - v)({}^ps_{11} + {}^ps_{12})] - 2\,kv({}^mq_{31})^2\}/$$
$$\{\mu_0[v^2\,({}^ps_{11} + {}^ps_{12}) + (1 - 2v)({}^ps_{11} + {}^ps_{12}) + kv(1 - v)({}^ms_{11} + {}^ms_{12})]$$
$$+ {}^m\mu_{33}\{v(1 - v)({}^ps_{11} + {}^ps_{12}) + kv^2\,({}^ms_{11} + {}^ms_{12}) - 2kv^2\,({}^mq_{31})^2\} \quad (2.16)$$

$$\alpha_{33} = 2\frac{k\mu_0(v - 1)\,{}^pd_{31}\,{}^mq_{31}}{[\mu_0(v - 1) - {}^m\mu_{33}v][kv({}^ms_{12} + {}^ms_{11}) - ({}^ps_{11} + {}^ps_{12})(v - 1)] + 2\,{}^mq_{31}^2\,kv^2},$$
$$(2.17)$$

$$\alpha_{E,33} = \frac{E_3}{H_3} = 2\frac{\mu_0 kv(1-v)\,{}^pd_{31}\,{}^mq_{31}}{\{2\,{}^pd_{31}^2(1-v) + {}^p\varepsilon_{33}[({}^ps_{11} + {}^ps_{12})(v-1) - v({}^ms_{11} + {}^ms_{12})]\}} \times$$
$$\frac{[({}^ps_{11} + {}^ps_{12})(v-1) - kv({}^ms_{11} + {}^ms_{12})]}{\{[\mu_0(v-1) - {}^m\mu_{33}v][kv({}^ms_{12} + {}^ms_{11}) - ({}^ps_{11} + {}^ps_{12})(v-1)] + 2\,{}^mq_{31}^2\,kv^2\}}$$
$$(2.18)$$

The earlier expression obtained by Harshe *et al.* [6] matched to our theory for the special case of $k = 1$, provided that the magnetic field is applied only to the ferrite phase.

The model presented above allows for the determination of the longitudinal ME coefficients as functions of volume fractions, physical parameters of phases, and elastic–elastic interfacial coupling parameter k: as given by Eq. 2.18.

2.1.2.2 Transverse ME effect

This case corresponds to E and δE being applied along the X_3 direction, and H and δH along the X_1 direction (in the sample plane). The ME voltage coefficient is $\alpha'_{E,T} = \alpha'_{E,31} = \delta E_3/\delta H_1$. For this case, the non-zero components of the $^{p}s_{ij}$, $^{p}d_{ki}$, $^{m}s_{ij}$, $^{m}q_{ki}$, s_{ij}, d_{ki}, q_{ki}, α_{kn} tensors are summarized in Table 2.2.

Table 2.2 Non-zero coefficients of piezoelectric and magnetostrictive phases and homogeneous material for transverse field orientation.

Piezoelectric phase	
Piezoelectric coefficients	Compliance coefficients
$^{p}d_{15} = {}^{p}d_{24}$	$^{p}s_{11} = {}^{p}s_{22}$
$^{p}d_{31} = {}^{p}d_{32}$	$^{p}s_{12} = {}^{p}s_{21}$
$^{p}d_{33}$	$^{p}s_{13} = {}^{p}s_{23} = {}^{p}s_{31} = {}^{p}s_{32}$
	$^{p}s_{33}$
	$^{p}s_{44} = {}^{p}s_{66}$
	$^{p}s_{66} = 2\,({}^{p}s_{11} + {}^{p}s_{12})$

Magnetostrictive phase	
Piezomagnetic coefficients	Compliance coefficients
$^{m}q_{35} = {}^{m}q_{26}$	$^{m}s_{11} = {}^{m}s_{22} = {}^{m}s_{33}$
$^{m}q_{12} = {}^{m}q_{13}$	$^{m}s_{12} = {}^{m}s_{21} = {}^{m}s_{13} = {}^{m}s_{23} = {}^{m}s_{31} = {}^{m}s_{32}$
$^{m}q_{11}$	$^{m}s_{44} = {}^{m}s_{55} = {}^{m}s_{66}$

Homogeneous material		
Piezoelectric coefficients	Piezomagnetic coefficients	Compliance coefficients
$d_{15} = d_{24}$	$q_{35};\ q_{26}$	$^{p}s_{11};\ {}^{p}s_{22};\ {}^{p}s_{33}$
$d_{31} = d_{32}$	$q_{12};\ q_{13}$	$^{p}s_{12} = {}^{p}s_{21}$
d_{33}	q_{11}	$^{p}s_{13} = {}^{p}s_{31};\ {}^{p}s_{23} = {}^{p}s_{32}$
		$^{p}s_{44};\ {}^{p}s_{55};\ {}^{p}s_{66}$

The expressions for the effective permittivity, ME susceptibility, and transverse ME voltage coefficient are then respectively

$$\varepsilon_{33} = \frac{{}^{p}d_{31}^{2}(v-1) + {}^{p}\varepsilon_{33}\left[\left({}^{p}s_{11} + {}^{p}s_{12}\right)(1-v) + kv\left({}^{m}s_{11} + {}^{m}s_{12}\right)\right]}{v\left[\left({}^{p}s_{11} + {}^{p}s_{12}\right)(1-v) + kv\left({}^{m}s_{11} + {}^{m}s_{12}\right)\right]},$$

(2.19)

$$\alpha_{31} = \frac{(v-1)v\left({}^{m}q_{11} + {}^{m}q_{21}\right){}^{p}d_{31}k}{(v-1)\left({}^{p}s_{11} + {}^{p}s_{12}\right) - kv\left({}^{m}s_{11} + {}^{m}s_{12}\right)},$$

(2.20)

$$\alpha_{E,31} = \frac{E_3}{H_1}$$

$$= \frac{-kv(1-v)\left({}^{m}q_{11} + {}^{m}q_{21}\right){}^{p}d_{31}}{{}^{p}\varepsilon_{33}\left({}^{m}s_{12} + {}^{m}s_{11}\right)kv + {}^{p}\varepsilon_{33}\left({}^{p}s_{11} + {}^{p}s_{12}\right)(1-v) - 2k\,{}^{p}d_{31}^{2}(1-v)}.$$

(2.21)

2.1.2.3 In-plane longitudinal ME effect

Finally, we consider a bilayer laminate that is poled with an electric field E in the plane of the sample. We suppose that the in-plane fields H and δH are parallel, and that the induced electric field δE is measured in the same direction (i.e., along the c-axis). The ME voltage coefficient is then defined as $\alpha'_{E,IL} = \alpha'_{E,11} = \delta E_1/\delta H_1$. Expressions for the effective parameters ε, μ, a, and a'_E can be obtained from Eqs. 2.7–2.13. These expressions are provided in Eqs. 2.22 and 2.23, and subsequently summarized in Table 2.3.

$$\varepsilon_{11} = {}^{m}\varepsilon_{11}(1-v) + {}^{p}\varepsilon_{11}v + \{k^2v^2(1-v)[{}^{p}d_{12}^{2}\,{}^{m}s_{11} + 2\,{}^{p}d_{11}^{2}\,{}^{m}s_{11}\,{}^{m}s_{12}\,{}^{p}d_{12}\,{}^{p}d_{11}$$

$$- vk(1-v)^2(2\,{}^{p}s_{12}\,{}^{p}d_{12}\,{}^{p}d_{11} - {}^{p}d_{11}^{2}\,{}^{p}s_{33} - {}^{p}d_{12}^{2}\,{}^{p}s_{11})]\}/$$

$$[({}^{p}s_{33}\,{}^{p}s_{11} - {}^{p}s_{12}^{2})(1-v)^2 + k^2v^2({}^{m}s_{11}^{2} - {}^{m}s_{12}^{2})$$

$$+ ({}^{p}s_{33}\,{}^{m}s_{11} + {}^{m}s_{11}\,{}^{p}s_{11} - 2\,{}^{m}s_{12}\,{}^{p}s_{12})\,kv(1-v)], \qquad (2.22)$$

$$\mu_{11} = \mu_{11}(1-v) + v\mu_0 + vk(1-v)[(-{}^{m}q_{11}^{2}\,{}^{m}s_{11} - {}^{m}q_{12}^{2}\,{}^{m}s_{11} + 2\,{}^{m}q_{11}\,{}^{m}q_{12}\,{}^{m}s_{12})vk$$

$$- (1-v)({}^{m}q_{11}^{2}\,{}^{p}s_{33} + {}^{m}q_{12}^{2}\,{}^{p}s_{11} - 2\,{}^{m}q_{12}\,{}^{m}q_{11}\,{}^{p}s_{12})]/$$

$$[({}^{p}s_{11}\,{}^{p}s_{33} - {}^{p}s_{23}^{2})(1-v)^2 + k^2v^2({}^{m}s_{11}^{2} - {}^{m}s_{12}^{2})$$

$$+ ({}^{p}s_{33}\,{}^{m}s_{11} + {}^{m}s_{11}\,{}^{p}s_{11} - 2\,{}^{m}s_{12}\,{}^{p}s_{12})\,kv(1-v)], \qquad (2.23)$$

$$a_{11} = \{[{}^m q_{11}({}^p s_{33}{}^p d_{11} - {}^p s_{12}{}^p d_{12}) + {}^m q_{12}({}^p s_{11}{}^p d_{12} - {}^p s_{12}{}^p d_{11})](1-v)$$
$$+ [{}^m q_{11}({}^m s_{11}{}^p d_{11} - {}^m s_{12}{}^p d_{12}) + {}^m q_{12}({}^m s_{11}{}^p d_{12} - {}^m s_{12}{}^p d_{11})]vk\}vk$$
$$(1-v)/[({}^p s_{33}{}^p s_{11} - {}^p s_{12}^2)(1-v)^2 + k^2 v^2 ({}^m s_{11}^2 - {}^m s_{12}^2)$$
$$+ ({}^p s_{33}{}^m s_{11} + {}^m s_{11}{}^p s_{11} - 2{}^m s_{12}{}^p s_{13})kv(1-v)], \qquad (2.24)$$

$$\alpha_{E,31} = (({}^m q_{11}({}^p s_{33}{}^p d_{11} - {}^p s_{12}{}^p d_{12}) + {}^m q_{12}({}^p s_{11}{}^p d_{12} - {}^p s_{12}{}^p d_{11}))(1-v)$$
$$+ ({}^m q_{11}({}^m s_{11}{}^p d_{11} - {}^m s_{12}{}^p d_{12}) + {}^m q_{12}({}^m s_{11}{}^p d_{12} - {}^m s_{12}{}^p d_{11}))vk)vk\,(1-v)/$$
$$(((1-p){}^m \varepsilon_{11} + v^p \varepsilon_{11})((1-v)^2({}^p s_{11}{}^p s_{33} - {}^p s_{12}^2) + (1-v)vk$$
$$({}^m s_{11}{}^p s_{11} + {}^p s_{33}{}^m s_{11} - 2\,{}^p s_{12}{}^m s_{12}) + k^2 v^2 ({}^m s_{11}{}^2 - {}^m s_{12}^2))$$
$$- kv\,(1-p)^2(2\,{}^p s_{12}{}^p d_{11}{}^p d_{12} - {}^p s_{33}{}^p d_{11}^2 - {}^p s_{11}{}^p d_{12}^2$$
$$+ k^2 v^2 (1-v)\,({}^m s_{11}{}^p d_{12}^2 + {}^m s_{11}{}^p d_{11}^2 - 2{}^m s_{12}{}^p d_{11}{}^p d_{11}). \qquad (2.25)$$

Amongst all the cases presented so far, the in-plane ME coefficient is expected to be the largest. This is due to availability of magnetostrictive and piezoelectric phases with high q- and d-values, respectively; and, to the absence of demagnetization fields. We will further use these outcomes later in the estimation of ME parameters for some specific examples.

Table 2.3 Non-zero coefficients of piezoelectric and magnetostrictive phases and homogeneous material for in-plane longitudinal field orientation.

Piezoelectric phase	
Piezoelectric coefficients	Compliance coefficients
${}^p d_{35} = {}^p d_{26}$	${}^p s_{33} = {}^p s_{22}$
${}^p d_{13} = {}^p d_{12}$	${}^p s_{32} = {}^p s_{23}$
${}^p d_{11}$	${}^p s_{13} = {}^p s_{23} = {}^p s_{31} = {}^p s_{32}$
	${}^p s_{11}$
	${}^p s_{66} = {}^p s_{55}$
	${}^p s_{44} = 2\,({}^p s_{33} + {}^p s_{32})$
Magnetostrictive phase	
Piezomagnetic coefficients	Compliance coefficients
${}^m q_{35} = {}^m q_{26}$	${}^m s_{11} = {}^m s_{22} = {}^m s_{33}$
${}^m q_{13} = {}^m q_{12}$	${}^m s_{12} = {}^m s_{21} = {}^m s_{13} = {}^m s_{23} = {}^m s_{31} = {}^m s_{32}$
${}^m q_{11}$	${}^m s_{44} = {}^m s_{55} = {}^m s_{66}$

Homogeneous material		
Piezoelectric coefficients	Piezomagnetic coefficients	Compliance coefficients
$^{p}d_{3}; {}^{p}d_{26}$	$^{m}q_{35}; {}^{m}q_{26}$	$^{p}s_{11}; {}^{p}s_{22}; {}^{p}s_{33}$
$^{p}d_{13}; {}^{p}d_{12}$	$^{m}q_{13}; {}^{m}q_{12}$	$^{p}s_{12} = {}^{p}s_{21}$
$^{p}d_{11}$	$^{m}q_{11}$	$^{p}s_{13} = {}^{p}s_{31}; {}^{p}s_{23} = {}^{p}s_{32}$
		$^{p}s_{44}; {}^{p}s_{55}; {}^{p}s_{66}$

2.1.3 ME Effect in Clamped Samples

The theory of ME coupling for laminates that are free from any exterior mechanical forces has been considered above. Next, please consider the same laminates when clamped in the direction axis 3 [10, 12, 30]. Clamping changes the character of the mechanical connection of phases in the laminates, which then significant modifies ME exchange between layers.

2.1.3.1 Longitudinal ME effect

The boundary conditions for the free case were $T_1 = T_2 = T_3 = 0$. However, in the clamped state, the same boundary conditions become $T_1 = T_2 = 0$ and $S_3 = s_{c33}T_3$, where $s_{c33} = S_3/T_3$ describes the compliance of the clamped system. For free samples $s_{c33} \gg s_{33}$, and for rigidly clamped ones $s_{c33} = 0$.

In this case the longitudinal ME voltage coefficient is determined by the expression

$$\alpha'_{E,33} = \frac{\alpha_{33}\left(s_{33} + s_{c33}\right) - d_{33}q_{33}H}{\varepsilon_{33}\left(s_{33} + s_{c33}\right) - d_{33}^{3}},\qquad (2.26)$$

where

$$s_{33} = \left(\left(v(1-v)2k\left({}^{m}s_{12} - {}^{p}s_{13}\right)^{2} - {}^{p}s_{33}\left({}^{p}s_{11} - {}^{p}s_{12}\right) - k{}^{m}s_{11}\left({}^{m}s_{11} + {}^{m}s_{12}\right)\right)$$
$$+ {}^{m}s_{11}\left({}^{p}s_{11} + {}^{p}s_{12}\right)\left(2v-1\right) - v^{2}k{}^{p}s_{33}\left({}^{m}s_{11} + {}^{m}s_{12}\right) - v^{2m}s_{11}\left({}^{p}s_{11} + {}^{p}s_{12}\right)\right)/$$
$$\left(\left({}^{p}s_{12} + {}^{p}s_{11}\right)(v-1) - kv\left({}^{m}s_{11} + {}^{m}s_{12}\right),\qquad (2.27)$$

$$d_{33} = \frac{2{}^{p}d_{31}k(v-1)\left({}^{m}s_{12} + {}^{p}s_{13}\right) + {}^{p}d_{33}\left(\left({}^{p}s_{11} + {}^{p}s_{12}\right)(v-1) - kv\left({}^{m}s_{11} + {}^{m}s_{12}\right)\right)}{\left({}^{p}s_{12} + {}^{p}s_{11}\right)(v-1) - kv\left({}^{m}s_{11} + {}^{m}s_{12}\right)^{2}},$$

$$(2.28)$$

$$q_{33} = \frac{\mu_0(1-v)(2kv\,^mq_{31}\left(^Ps_{13}-^ms_{12}\right)+^mq_{33}((1-v)(^Ps_{11}+^Ps_{12})+kv(^ms_{11}+^ms_{12})))}{((1-v)\mu_0+^m\mu_{33}v)(kv\left(^ms_{11}+^ms_{12}\right)+(1-v)\left(^Ps_{11}+^Ps_{12}\right)-2k\,^mq_{31}{}^2v^2}.$$

(2.29)

2.1.3.2 Transverse ME effect

In the clamped condition ($T_1 = T_2 = 0$ and $S_3 = s_{c33}T_3$), the transverse ME voltage coefficient is given by

$$\alpha'_{E.31} = \frac{\alpha_{31}(s_{33}+s_{A33})-d_{33}q_{13}}{\varepsilon_{33}(s_{33}+s_{A33})-d_{33}{}^2},$$

(2.30)

where s_{33} and d_{33} are defined by Eqs. 2.27 and 2.28, and

$$q_{13} = -\{v^{2m}q_{12}\,[(^Ps_{11}+^Ps_{12})-k(^Ps_{13}+^ms_{11})]+^mq_{12}\,(1-2v)$$
$$(^Ps_{11}+^Ps_{12})+kv^mq_{12}\,(^Ps_{13}+^ms_{11})+kv(1-v)^mq_{11}\,(^Ps_{13}-^ms_{12})\}/$$
$$[(^Ps_{12}+^Ps_{11})\,(v-1)-kv(^ms_{11}+^ms_{12})\,].$$

(2.31)

2.1.3.3 In-plane longitudinal ME effect

In the clamped condition ($T_1 = T_2 = 0$ and $S_3 = s_{c33}T_3$), the in-plane longitudinal ME voltage coefficient is given as

$$\alpha'_{E,11} = -\frac{\alpha_{11}(s_{33}+s_{A33})-d_{13}q_{13}}{\varepsilon_{33}(s_{33}+s_{A33})-d_{13}{}^2},$$

(2.32)

where s_{33} and q_{13} are defined by Eqs. 2.27 and 2.31, and

$$d_{13} = \{^Pd_{13}v\{k^Ps_{13}\,[^Ps_{11}\,(v-1)^2+vk^ms_{11}\,(1-v)]+^ms_{12}k[kv[^ms_{12}$$
$$-(^Ps_{13}+^ms_{11})(1-v)]-^Ps_{11}\,(1-v)^2+^Ps_{13}\,(1-v^2+kv^2)]-^Ps_{11}\,^Ps_{33}$$
$$(v-1)^2-^Ps_{13}^2\,(v-1)^2\,(k-1)+kv(v-1)\,^ms_{11}\,(^Ps_{11}+^Ps_{33})$$
$$-k^2v^2\,^ms_{11}^2\}+^Pd_{11}\,vk\{^ms_{12}\,v[(^ms_{12}-^Ps_{13}-^ms_{11})\,k(1-v)$$
$$-^Ps_{33}\,(v-2+k)]-(-1+v)\,^Ps_{13}\,[kv\,^ms_{11}+(^ms_{12}+^Ps_{33}-^Ps_{13})$$
$$(1-v)]\}\}/[(1-v)^2\,(^Ps_{13}^2-^Ps_{11}\,^Ps_{33})+k^2v^2\,(^ms_{12}^2-^ms_{11}^2)$$
$$-v\,(1-v)\,k\,(^ms_{11}\,(^Ps_{11}+^Ps_{33})-2\,^Ps_{13}\,^ms_{12})].$$

(2.33)

It is interesting to consider two limiting cases: free and rigidly clamped samples. The first case corresponds to the condition that $s_{c33} \gg s_{33}$ in which case Eqs. 2.26, 2.30, and 2.32 reduce respectively to Eqs. 2.18, 2.21, and 2.25. However, for rigidly clamped samples (i.e., $s_{c33} = 0$), we obtain the following expressions.

$$\alpha'_{E,33} = -\frac{\alpha_{33}s_{33} - d_{33}q_{33}}{\varepsilon_{33}s_{33} - d_{33}^2}, \tag{2.34}$$

$$\alpha'_{E,31} = -\frac{\alpha_{31}s_{33} - d_{33}q_{13}}{\varepsilon_{33}s_{33} - d_{33}^2}, \tag{2.35}$$

$$\alpha'_{E,11} = -\frac{\alpha_{21}s_{33} - d_{13}q_{13}}{\varepsilon_{13}s_{33} - d_{13}^2}, \tag{2.36}$$

2.1.4 Examples of Multilayer Structures

The preceding comprehensive theoretical treatment resulted in expressions of the ME voltage coefficients for three different orientations of fields, which were the ones of most importance, including: longitudinal, transverse, and in-plane longitudinal. The most significant features of the model are as follows: (i) Consideration of three different field configurations for two different sample conditions, which are unclamped and clamped. This allows for the determination of a single-valued interface parameter k, facilitating quantitative characterization of the bilayer interface. (ii) Consideration of a new field configuration, i.e., in-plane longitudinal fields that has very strong ME coupling. And, (iii) consideration of the effect of a finite magnetic permeability on the magnetostriction of the magnetic subsystem: which was ignored in prior investigations.

Next, we extend the theoretical treatment to the case of ME coupling in multi-layered composites. Consider the materials couple cobalt ferrite and barium titanate (CFO–BTO), which is a system that has been of significant prior interests [4]. Since the value of α_E depends notably on the concentration of the two phases, the longitudinal and transverse voltage coefficients have been determined as a function of the volume fraction v of the piezoelectric phase in CFO–BTO.

Results of calculations using the model are illustrated in Fig. 2.2, which were obtained by assuming an ideal interface coupling ($k = 1$) and by using material parameters given in Table 2.4.

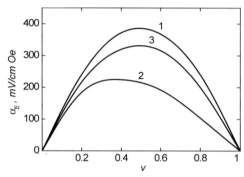

Figure 2.2 Transverse magnetoelectric (ME) voltage coefficient $\alpha_{E,31} = \delta E_3/\delta H_1$ and longitudinal coefficient $\alpha_{E,33} = \delta E_3/\delta H_3$ for a perfectly bonded ($k = 1$) bilayer structure consisting of CFO and barium titanate (BTO) (1 is the transverse orientation, 2 the longitudinal, and 3 the direction of the coefficient $^pE_3/^mH_3$ [11]).

Let us now consider the v-dependence of $\alpha_{E,33}$. Calculations using the theoretical treatment are shown in Fig. 2.2. ME coupling is obviously absent in pure CFO and BTO phases. As v is increased from 0, $\alpha_{E,33}$ increases, approaching a maximum near $v_m = 0.4$. The coupling then weakens with further increase in v, becoming zero at $v = 1$. Prior theoretical studies of $\alpha_{E,33}$ by Harshe *et al.* [6] reported values that were 30–40% higher than ones presented here, also shown in Fig. 2.2, which is due to the implied assumption that $\mu_{33} = \mu_0$. Obviously, demagnetizing fields associated with the longitudinal orientation result in a reduction of $\alpha_{E,33}$.

Table 2.4 Material parameters–compliance coefficient s, piezomagnetic coupling q, piezoelectric coefficient d, permeability ε, and permittivity μ for lead zirconate titanate (PZT), cobalt ferrite (CFO), barium titanate (BTO), nickel ferrite (NFO), and lanthanum strontium manganite [2, 3].

Material	s_{11} (10^{-12} m²/N)	s_{12} (10^{-12} m²/N)	s_{13} (10^{-12} m²/N)	s_{33} (10^{-12} m²/N)	q_{33} (10^{-12} m/A)	q_{31} (10^{-12} m/A)	d_{31} (10^{-12} m/V)	d_{33} (10^{-12} m/V)	μ_{33}/μ_0	$\varepsilon_{33}/\varepsilon_0$
PZT	15.3	−5	−7.22	17.3			−175	400	1	1750
BTO	7.3	−3.2					−78		1	1345
CFO	6.5	−2.4			−1880	556			2	10
NFO	6.5	−2.4			−680	125			3	10
LSMO	15	−5			250	−120			3	10

For transversely applied fields, one observes similar features as shown in Fig. 2.2, but the maximum in $\alpha_{E,31}$ is almost a factor of two higher than that of $\alpha_{E,33}$. The transverse coefficient peaks at a slightly higher value of v, relative to that of the longitudinal case. The key findings of Fig. 2.2 are predictions (i) of a giant ME effect in CFO–BTO, (ii) that the transverse ME coupling is much stronger than the longitudinal one, and (iii) that a maximum in both transverse and longitudinal ME coupling occurs near equal volume fractions of constituent phases. The values predicted by the model greatly exceed those previously found in single phase ME crystals — by orders of magnitude: thus, the term giant is designated to the ME effect in laminate composites to distinguish this fact.

The bilayer all-ceramic composite of importance is CFO–PZT (lead zirconate titanate): its longitudinal ME coefficient and subsequent dependence on k are given in Table 2.4. The variation of $\alpha_{E,33}$ with v for various values of k is shown in Fig. 2.3a. The magnitude of $\alpha_{E,33}$ decreases with decreasing k, and v_{max} shifts to PZT-rich compositions. Figure 2.3b shows the dependence of the maximum value in $\alpha_{E,33}$ on k, where calculations are illustrated for various values of v_{max}. With increasing k, a near-linear increase was found in the maximum value of $\alpha_{E,33}$.

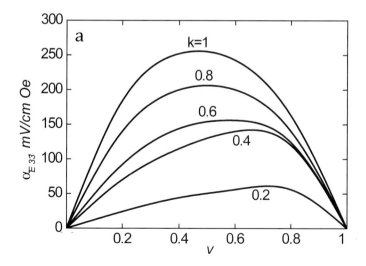

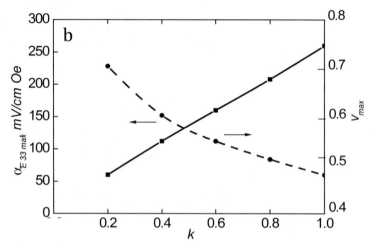

Figure 2.3 (a) Estimated dependence of longitudinal ME voltage coefficient on interface coupling k and volume fraction v for CFO–lead zirconate titanate (PZT) bilayer. (b) Variation with k of maximum $\alpha_{E,33}$ and the corresponding v_{max}.

Next, let us consider the ME effect in CFO–PZT for the two other field orientations: transverse and in-plane longitudinal [13, 14]. The v-dependence of α_E for both cases is shown in Fig. 2.4, assuming an ideal interface coupling ($k = 1$). The insets show variations in $\alpha_{E,max}$ and v_{max} with k. For transverse fields, the maximum α_E is 40% higher than that of $\alpha_{E,33}$. This is due to the strong parallel piezomagnetic coupling q_{11} which determines $\alpha_{E'}$, relative to that of q_{31} which determines $\alpha_{E,33}$. The transverse ME coupling has a higher value of v_{max}, compared to the longitudinal one. Another notable feature is that $\alpha_{E',max}$ vs k and v_{max} vs k behave similarly for both longitudinal and transverse cases.

The most significant prediction of the present model is that the strongest ME coupling should occur for in-plane longitudinal fields, as shown in Fig. 2.4b. One can easily see in Fig. 2.4 that when the field is switched from longitudinal to in-plane longitudinal that the maximum value of the relevant ME coefficient increases by nearly an order of magnitude: $\alpha_{E,max} = 260$ mV/(cmOe) for the longitudinal orientation, whereas $\alpha_{E,11} = 3600$ mV/(cmOe) for the in-plane longitudinal. The v-dependence of $\alpha_{E,11}$ is also shown in

Fig. 2.4b, which reveals a rapid increase in the ME coefficient to a maximum value of $\alpha_{E,11}$ = 3600 mV/(cmOe) for v = 0.11, which is followed by a near-linear decrease with further increase of v. Such an enhancement in the in-plane longitudinal coefficient relative to the longitudinal one is understandable due to (i) the absence of demagnetizing fields in the in-plane configuration, and (ii) increased piezoelectric and piezomagnetic coupling coefficients compared to longitudinal fields. The down-shift in the value of v_{max} (from 0.5 to 0.6 for longitudinal and transverse fields to a much smaller value of 0.1) is due to the concentration dependence of the effective permittivity, which can be determined by Eq. 2.19.

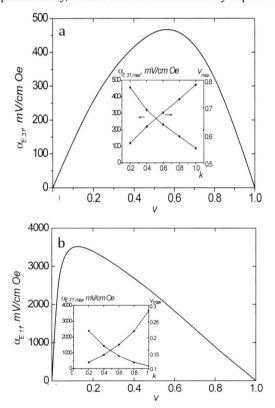

Figure 2.4 Results as in Fig. 2.2 for CFO–PZT bilayers but for (a) transverse fields ($\alpha_{E,31}$) and (b) in-plane longitudinal fields ($\alpha_{E,11}$). For in-plane longitudinal fields, the poling field and other dc and ac fields are parallel to each other and in the sample plane.

Consider now the effect of clamping on a bilayer of CFO–PZT [13]. Very significant changes in the nature of ME coupling are expected when the bilayer is subjected to a uniform stress applied perpendicular to the sample plane. Representative results for longitudinal fields are shown in Fig. 2.5. Variations in $\alpha'_{E,33}$ with the compliance parameter s_{c33} is depicted in Fig. 2.5a for a series of v-values. The results are for an interface coupling of $k = 0$. One can expect a value of $\alpha'_{E,33} = 0$ when $k = 0$ in unclamped ($s_{c33} = \infty$) bilayers. As the uniaxial stress on the laminate is increased (i.e., s_{c33} is decreased), one would expect $\alpha'_{E,33}$ to increase, reaching a peak value for zero compliance. This point of zero compliance corresponds to the case of rigidly clamped samples. The enhancement is rather large in laminates having equal volume fractions of the two phases (i.e., $v = 0.5$). Figure 2.5b shows estimated $\alpha'_{E,33}$ vs v for rigidly clamped CFO–PZT for a series of k-values.

It is quite intriguing that clamping associated enhancement in ME coupling is very high in samples with smaller interface coupling parameters k. Key inferences from Figs. 2.3 and 2.4 that can be made then are: (i) clamping, in general, leads to an increase in $\alpha'_{E,33}$; and (ii) that the largest increase occurs for rigidly clamped samples, with the smallest values of the interface coupling parameter k.

Similar results of clamping related effects on α'_E for transverse and in-plane longitudinal fields are shown in Fig. 2.6. One expects the highest α'_E for $k = 0$. Estimates of α'_E as a function of v are shown for representative k-values, and for rigidly clamped ($s_{c33} = 0$) bilayers. For transverse fields, a substantial reduction in $\alpha'_{E,31}$ can be seen in clamped samples, compared to that of unclamped samples (Fig. 2.4). The overall effect of clamping is a reduction in the strength of transverse ME voltage coefficient. Finally, Fig. 2.5b shows that for a clamped bilayer subjected to in-plane longitudinal fields that clamping related changes in the ME coefficient are quite weak compared to either the longitudinal or transverse ones.

Another bilayer of importance is nickel ferrite (NFO)–PZT. Although NFO is a soft ferrite with a much smaller anisotropy and magnetostriction than CFO, efficient magneto-mechanical coupling in NFO–PZT gives rise to ME voltage coefficients comparable to those of CFO–PZT. Using the model presented in this chapter, we

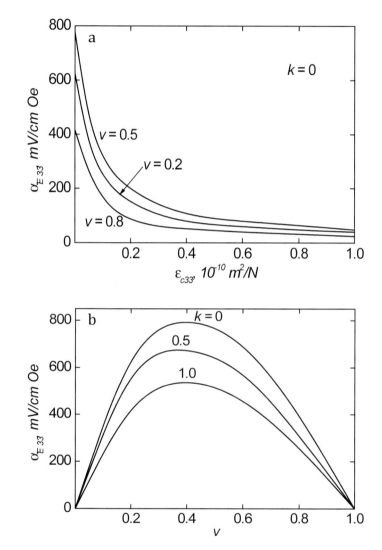

Figure 2.5 (a) Variation of the longitudinal ME voltage coefficient for CFO–PZT with the compliance coefficient s_{c33} for the clamping system. An infinite compliance of clamping corresponds to unclamped samples and zero compliance represents a rigidly clamped bilayer. Values are for $k = 0$ and for a series of volume fraction v for PZT. (b) $\alpha'_{E,33}$ vs v as a function of k for a rigidly clamped CFO–PZT.

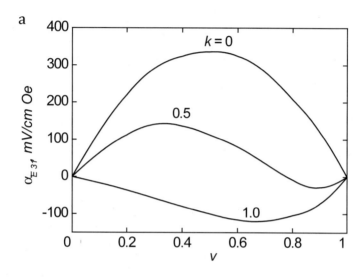

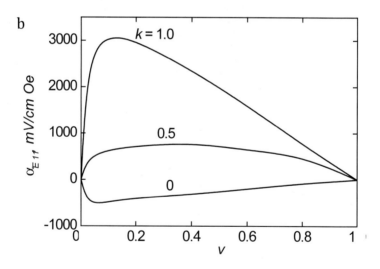

Figure 2.6 Dependence of transverse and in-plane longitudinal ME voltage coefficients on volume fraction v and interface coupling k for rigidly clamped (compliance $s_{c33} = 0$) CFO–PZT bilayer. Negative values for α'_E represent a 180° phase difference between δE and δH.

can estimate α_E for NFO–PZT for various field orientations and conditions, similar to that done above for CFO–PZT. Representative results for unclamped and rigidly clamped samples with ideal interface couplings ($k = 1$) are presented in Fig. 2.7. Key inferences that can be made from the results are as follows. (i) For unclamped bilayers, α_E is the smallest for longitudinal fields and is the highest for in-plane longitudinal fields. (ii) $\alpha_{E,31}$ and $\alpha_{E,11}$ are higher than $\alpha_{E,33}$ by a coefficient of 5–30×. (iii) Upon rigidly clamping a bilayer, there is a five-fold increase in $\alpha'_{E,33}$, a 50% reduction in $\alpha'_{E,31}$, and a very small decrease in $\alpha'_{E,11}$.

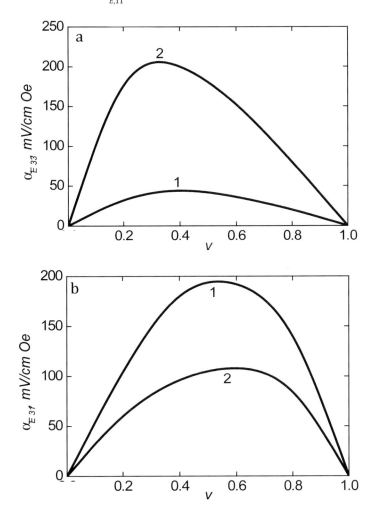

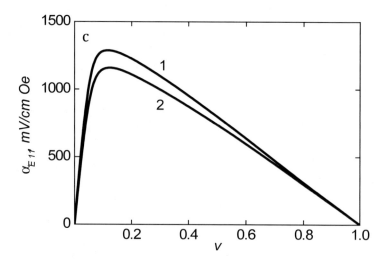

Figure 2.7 Concentration dependence of longitudinal, transverse, and in-plane longitudinal ME voltage coefficients for unclamped and rigidly clamped nickel ferrite (NFO)–PZT bilayer for interface coupling $k = 1$.

Finally, we consider composites that have lanthanum strontium manganites for the magnetostrictive phase. Lanthanum manganites with divalent substitutions have attracted considerable interest in recent years due to double exchange mediated ferromagnetism, metallic conductivity, and giant magnetoresistance [16]. The manganites are potential candidates for ME composites because of (i) high magnetostriction and (ii) metallic conductivity that eliminates the need for a foreign electrode at the interface. Figure 2.8 shows the longitudinal and transverse ME voltage coefficients for unclamped $La_{0.3}Sr_{0.7}MnO_3$ (LSMO)–PZT bilayers that assumes ideal coupling at the interface. In this case, the values of the ME coefficients are quite small compared to that of ferrite–PZT: this is due to weak piezomagnetic coefficients and compliances parameters for LSMO. The ME coefficient for in-plane longitudinal fields and the effects of clamping for different field orientations were similar in nature to those for ferrite–PZT bilayers, and thus are not discussed in any detail here.

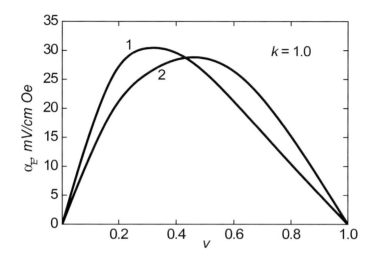

Figure 2.8 (1) Longitudinal and (2) transverse ME voltage coefficients as a function of volume fraction v for PZT in unclamped $La_{0.3}Sr_{0.7}MnO_3$ (LSMO)–PZT bilayer for interface coupling $k = 1$.

2.1.5 Experimental Data

It is important to compare the theoretical predictions, illustrated in Figs. 2.2–2.8, with experimental data. There have been prior investigations of ME layered composites, both for cases of longitudinal and transverse effects on unclamped samples. To our knowledge, such measurements have never been done for in-plane longitudinal fields. We also discuss recent data for clamped NFO–PZT. Other results that will be cited include those obtained on bilayer or multilayer composites processed either by high temperature sintering, or by epoxying thick films/disks of ferrite/manganite–PZT. Samples were poled in an electric field, and the ME coefficient measured by subjecting the sample to a dc magnetic bias H and an ac magnetic field δH while measuring the ac induced electrical field.

A typical measurement system for the ME effect is illustrated in Fig. 2.9 [8, 15, 28]. It consists of a dc electromagnet, between which is located Helmholtz coils. Samples to be measured are then placed between the Helmoltz coils. The electromagnet should be driven by a constant-current source, and the Helmholtz coils are typically driven by a low-frequency oscillator with a frequency bandwidth of $1 < f < 10^5$ Hz. To measure the induced voltage, the electrodes of a sample are then electrically connected to an operational amplifier, which subsequently feeds into either an oscilloscope or a lock-in amplifier.

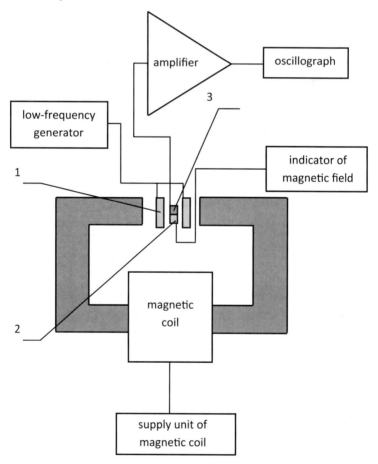

Figure 2.9 Measuring scheme for research of magnetoelectric effect in ferrite–piezoelectric composites: 1 — coils of Helmholtz, 2 — a probe of the gauss-meter, 3 — a researched material.

The operational amplifier needs to have an input impedance (>100 MΩ) that is significantly larger than that of the oscilloscope (~1 MΩ), to have accurate measurements: this will ensure a maximum signal gain of 60 dB. To read the voltage by this method, we need to consider Fig. 2.10 that illustrates the input impedance of the amplifier and an internal resistance of the sample under testing. The voltage induced across the sample in figure is

$$U_0 = \frac{U_{out}(R_s + R_{in})}{R_{in}} \tag{2.37}$$

where, U_{out} is the measured voltage in the circuit of Fig. 2.10; R_s is the resistance of the sample; R_{in} is the input impedance of the operational amplifier; and U_0 is the actual voltage on the sample.

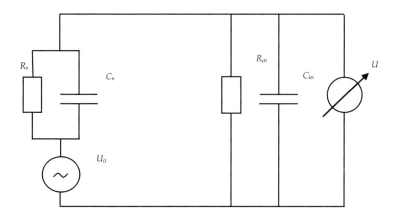

Figure 2.10 The equivalent circuit of measuring set.

The corresponding dielectric field intensity (E) induced on the sample can be calculated from the well-known relationship:

$$E = \frac{U_0}{d}, \tag{2.37a}$$

where d is thickness of a material. Finally, the ME voltage coefficient can then be calculated by

$$a_E = \Delta E / \Delta H. \tag{2.38}$$

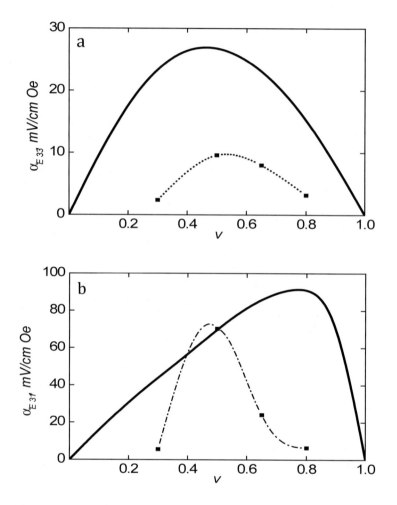

Figure 2.11 Concentration association longitudinal (a) transverse (b) ME coefficient on voltage of CFO–PZT compositions. Solid line — the theory for $k = 0.1$, points — experiment.

Let us consider first laminate composites of CFO–PZT. Figure 2.11 shows α_E for bilayers and sintered multilayers as a function of v. These data were taken at room temperature data taken at low frequencies (100–1000 Hz). The desired volume fractions v were achieved by careful control of the layer thickness. Data for both longitudinal and transverse fields show an increase in α_E with v until a maximum is reached, as earlier predicted by theory. However, these data clearly demonstrated that the actual experimental value is an order of magnitude smaller than that predicted in Figs. 2.3 and 2.4 (assuming $k = 1$). It is, therefore, logical to compare the data with calculated values of α_E as a function of v using a reduced interface coupling parameters of $k = 0.1$: in this case, agreement between theory and experiment can be seen, as shown in Fig. 2.11. The key inference that can be made concerns the inherently poor interface coupling for CFO–PZT, irrespective of sample synthesis techniques. We address possible causes for this poor coupling later in this section.

However, interestingly, a similar investigation for NFO–PZT indicates a near ideal interface coupling. Figure 2.12 shows experimental data for α_E as a function of v for transverse fields in the case of unclamped multilayers. These data taken from [17] are in agreement with theoretical ones, as shown in Fig. 2.12, calculated assuming an ideal interface parameter of $k = 1$: in particular, for $0.4 < v < 0.8$. Further evidence demonstrating an efficient stress mediated electromagnetic coupling in NFO–PZT is presented in Fig. 2.13. Data showing the dependence of $\alpha_{E,31}$ on dc magnetic bias is shown for both unclamped and rigidly clamped multilayers, assuming $v = 0.5$. In this figure, the ME coefficient can be seen to drop to zero at higher fields, when the magnetostriction attains saturation. The theoretical estimates given in Fig. 2.13 are for q-values obtained from λ vs H data reported in [17]. The excellent agreement between theory for $k = 1$ and the experimental data reinforces the ideal interface conditions inferred just above. Another important observation from Fig. 2.13 is the observation of a reduction in $\alpha'_{E,31}$ for a rigidly clamped sample, which is in agreement with theoretical predictions summarized in Fig. 2.7. Similar comparison of data and theory ($k = 1$) for longitudinal fields has been found for NFO–PZT, also an increase in $\alpha'_{E,33}$ for clamped samples has been observed as predicted by theory.

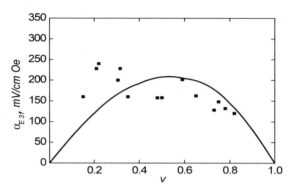

Figure 2.12 Volume fraction dependence of transverse ME voltage coefficient for NFO–PZT. Solid line — theory for $k = 0.1$, points — experiment.

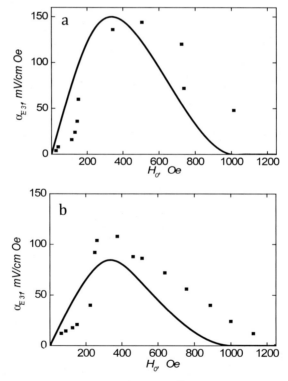

Figure 2.13 Transverse ME voltage coefficient vs H_0 and data for free (a) and rigidly clamped (b) NFO–PZT for $k = 1$: Solid line — theory, points — experiment.

A third materials couples, LSMO–PZT, is considered in Fig. 2.14, which shows α_E as a function of v for longitudinal and transverse fields for the cases of unclamped bilayers and multilayers. The α_E values are the smallest amongst the three systems considered here. Calculated values assuming $k = 0$ were found to be quite high compared to the data, rather it was found that nonideal values of $k = 0.2$ gave reasonable agreement with the data. Thus, one can readily infer poor interfacial coupling in LSMO–PZT, similar to that for CFO–PZT.

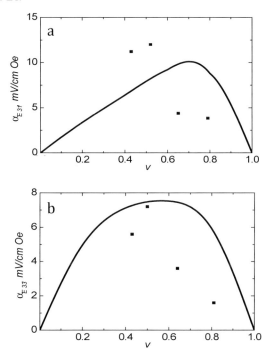

Figure 2.14 Volume fraction dependence of transverse (a) and longitudinal (b) ME voltage coefficients for LSMO–PZT: Solid line — the theory for $k = 0.2$, points — experiment.

Finally, we should comment on a possible cause of poor interfacial coupling for CFO–PZT and LSMO–PZT, and ideal coupling for NFO–PZT. The parameter k can be expected to be sensitive to mechanical, structural, chemical, and electromagnetic parameters at the interface. We attribute unfavorable interface conditions in CFO–PZT and LSMO–PZT to inefficient magneto-mechanical coupling. The magneto-mechanical coupling k_m is given by $k_m = (4\pi\lambda'\mu_r/E)^{1/2}$; where λ' is the dynamic magnetostrictive constant and μ_r is the reversible permeability, and E is Young's modulus. In ferrites, under the influence of a dc magnetic bias H and ac magnetic field δH, domain wall motion and domain rotation contribute to the Joule magnetostriction and consequently to the effective linear piezomagnetic coupling. A key requirement for strong coupling is unimpeded domain wall motion and domain rotation. A soft ferrite with a high initial permeability (i.e., low anisotropy), such as NFO, will have key materials parameters favoring a high k_m, and consequently, strong ME effects. Measurements have shown that NFO has an initial permeability of 20, whereas that of LSMO and CFO is 2–3. Thus, one can infer a plausible simple explanation of the near interfacial parameter for NFO–PZT is (in part) favorable domain motion.

2.2 BULK COMPOSITES

Design of new ME composites assumes the use of reliable theoretical models, allowing prediction of properties for various materials couples and over a range of laminate parameters such as v. The field of piezoelectric composites is quite mature. Manufacturing methods of all-ceramic composites are based on an initial mixing of starting powders batched in proportion to the composite volume fraction, followed by pressing and densification/sintering to a net-shape. Clearly, if the concentration of one of the constituent phases is small, then that phase will consist of isolated particles in a matrix. Following accepted classification nomenclature [5] this composite should be referred to as a 0–3 type, as one phase is isolated (i.e., connected in zero dimensions) and the second is interconnected in three dimensions. If the volume fraction of the secondary phase in the matrix is increased, and a percolation limit is reached, then it is classified as a 1–3 type composite. If the secondary phase then crosses that initial percolation limit, and subsequently begins to be interconnected in two dimensions, the composite connectivity is known as the 2–3 type. We mention these things at this time to make the point that the same ceramic manufacturing technology allow the fabrication of a wide range of relative volume fractions of the different phases in an all-ceramic composite, and consequently to various possible types of dimensional interconnectivities. Accordingly, it is very important to choose the correct method of calculation for effective constants of a composite at various relative volume fractions of components.

Unfortunately, exact solutions of three-dimensional problems related to the calculation of effective constants of inhomogeneous systems are unknown. Therefore, there is presently no precise structural classification of composites. Within the limited theory of heterogeneous systems of two-phase composites, there are two principle approaches to approximate solutions: matrix systems and two-component mixtures, for which behavior of effective parameters depending on concentration continuously.

In the case of matrix systems, modification of the concentration from 0 to 1 does not change the qualitatively structure of the composite: at any concentration, one of the components must form a coherent matrix that contains isolated particles of the second component. The system always remains essentially non-central, and matching formulas for an evaluation of effective constants give their continuous dependence on concentration in the entire range from 0 to 1. We should note that the application of these formulas to the calculation of effective constants of composites is not always justified.

The case of two-component mixtures is characterized by a qualitative modification of the structure of the composite, as the concentration is changed. Such systems are characterized, as is well known, by critical concentrations at which point there are important property changes such as metal–insulator or rigid-plastic transformations. The metal–insulator transformation occurs in a composite consisting of an insulating and conductive phases. Assume that the insulating phase is initially the matrix and that the conductive one consists of isolated particles. In this case, initially the composite is insulating; however, when the percolation limit is crossed, the conducting particles form an interconnected conduction pathway, dramatically lowering the resistivity near a critical volume fraction. In the second example (rigid-plastic transformation), it is supposed that the composite is a mixture in which the elastic compliance of one of the constituent phases tends to infinity (for example, a porous composite). This composite type posses a critical concentration of the second phase, above which the rigid framework of the composite loses its stability. It should be straightforward to see that any bulk composites will have numerous effective materials properties, all of which change with relative phase volume fraction from in a manner independent of other properties.

However, due to strictive couplings, the ME properties (the coupling between magnetization and polarization) are not independent. Rather, ME effects are a direct consequence of an interdependence — a product tensor property, introduced by the action of one phase on the other via strictions. To predict the properties of ME composites, an effective medium approach should be used.

Let us then consider a composite with a 3–0 type connectivity. In Fig. 2.15, a ME composite with a connectivity 3–0 is represented.

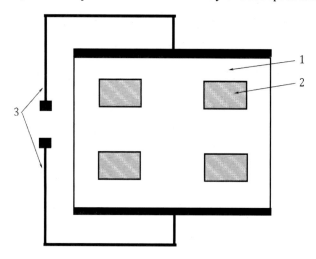

Figure 2.15 Composite with connectivity 3–0: 1 — piezoelectric phase; 2 — magnetostriction phase; 3 — electrodes.

Cubic models for ferrite–ferroelectric composites with a connectivity of 3–0 and 0–3 have been considered by Harshe [7]. Numerically, the ME coefficient is equal to the ratio of the electric field induced on the composite by an applied magnetic field: the ME coefficient is equal to E_3/H_3. It is necessary to realize that the magnetic field was applied only to the ferrite phase: i.e., $E_3/{}^m H_3$ where ${}^m H_3$ is the local magnetic field on the ferrite phases which may exceed that applied to the entire composite. Harshe's study only considered the case of free cubic cells, and effective parameters of the composite in known model systems were not determined. However, in real composites, we must consider the case of non-free cells. It is also very important to use any such model to predict the effective composite parameters. In the following section, we present a generalized model for ferrite–piezoelectric composites that allows one to define and predict the effective parameters of said composite using given conditions.

2.2.1 Connectivities 0–3 and 3–0

2.2.1.1 Bulk composite with connectivities 0–3

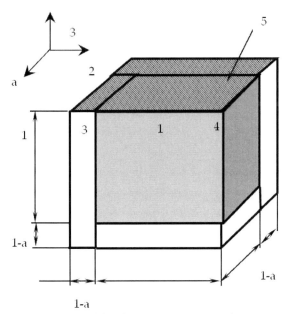

Figure 2.16   Cubic model of ME composite with connectivity 3–0. 1 — magnetostriction phase; 2-4 — piezoelectric phase; 5 — electrode.

The ME composite shown in Fig. 2.16 [7] has a 3–0 connectivity, where the piezoelectric phase has connectivity in all three directions of the Cartesian frame, and where the magnetostrictive phase is isolated. The properties of this ME composite will depend on the parameters of the corpuscles, and also on the terminal conditions [18–21, 29]. Now, let us suppose that the geometrical model for ME composites in this figure is miniaturized to fine scales.

If the given cubic model ME composite is considered as a material consisting of consecutive and parallel connections of cubic cells with legs of unit length, then it is obvious by the definition of properties of a composite that it is possible to consider only one cubic cell rather than the entire ensemble of cells. The magnetostriction phase (region 1 in Fig. 2.16) is enclosed by piezoelectric ones along different directions (regions 2, 3, 4). In Fig. 2.16, the magnetostriction phase is represented by a cube with side a; thus, the bulk magnetostriction phases has a volume of a^3, and the piezoelectric one of $(1 - a^3)$.

To derive the effective material parameters of a composite, we use an averaging method. Equations 2.7 and 2.8 for strain and dielectric displacement in the piezoelectric cells then become

$$^nS_i = {}^ns_{ij}\,{}^nT_j + {}^nd_{ki}\,{}^nE_k, \tag{2.39}$$

$$^nD_k = {}^nd_{ki}\,{}^nT_i + {}^n\varepsilon_{kn}\,{}^nE_n,$$

where $n = 2, 3, 4$ is the cube cell number; ^nS_i is the component of tensor strains of n-th cell; ^nE_k is the component of vector electric field intensity; ^nD_k is the component of vector electric displacement; ^nT_j is the stress tensor component; $^ns_{ij}$ is the coefficient of compliance; $^Pd_{ki}$ is the piezoelectric module; and $^P\varepsilon_{kn}$ is the tensor of permittivity.

Correspondingly, Eqs. 2.9 and 2.10 for the strain and magnetic induction of the piezomagnetic phase in cell 4 then becomes

$$^1S_i = {}^1s_{ij}\,{}^1T_j + {}^1q_{ki}\,{}^1H_k, \tag{2.41}$$

$$^1B_k = {}^1q_{ki}\,{}^1T_i + {}^1\mu_{kn}\,{}^1H_n, \tag{2.42}$$

where ^1S_i is the component of tensors strain magnetostriction phases; ^1T_j is the stress tensor component of magnetostriction phase; $^1s_{ij}$ is the coefficient of compliance; ^1H_k is the component of magnetic field vectors; ^1B_k is the component of magnetic displacement vectors; $^1q_{ki}$ is the piezomagnetic module; and $^1\mu_{kn}$ is the tensor of permeability.

Since the composite is considered as a homogeneous material characterized by effective parameters, the relations 2.11 and 2.12 given earlier for components of strains, electric, and magnetic inductions remain valid. The effective parameters of the composite can then be found by a joint solution of Eqs. 2.11, 2.12, and 2.39–2.42. To solve for specific equations, it is necessary to provide the boundary conditions for strains and stresses of composite cells. The boundary conditions for strains appearing in a cubic cell should be

$$^1S_1 = {}^2S_1, \tag{2.43}$$

$$^1S_2 - {}^2S_2 = 0, \tag{2.44}$$

$$^2S_1 - {}^3S_1 = 0, \tag{2.45}$$

$$^3S_3 - {}^4S_3 = 0, \tag{2.46}$$

$$a{}^1S_3 + (1-a)\,{}^2S_3 - {}^3S_3 = 0, \tag{2.47}$$

$$a{}^2S_2 + (1-a)\,{}^3S_2 - {}^4S_2 = 0. \tag{2.48}$$

It is possible to define the boundary conditions for stresses in the cells using balance conditions. There are numerous cases, which we now discuss. The sum of forces applied along direction 1 to cell 4 equals zero, given as

$$^1A_1{}^1T_1 + {}^2A_1{}^2T_1 + {}^3A_1{}^3T_1 = {}^4A_1{}^4T_1, \tag{2.49}$$

where ^iA_j is the sectional area of i-th cell which is perpendicular to axis j. The force applied along axis 1 to cube 4 also equals zero:

$$^4T_1 = 0. \tag{2.50}$$

The sum of the forces applied along axis 2 to cells 3 and 4 is also zero:

$$^3A_2{}^3T_2 + {}^4A_2{}^4T_2 = 0. \tag{2.51}$$

Owing to the sum of the forces applied to cell 3 along axis 2 being equal to the forces applied to cubes 1 and 2, we can obtain

$$^1A_2{}^1T_2 + {}^2A_2{}^2T_2 = {}^3A_2{}^3T_2. \tag{2.52}$$

Owing to the stress applied to cell 1 along axis 3 being equal to stress applied to cube 2 along axis 3, we obtain

$$^1T_3 - {}^2T_3 = 0. \tag{2.53}$$

The sum of forces applied to cells 2, 3, and 4 along axis 3 equals 0; consequently, we can find the expression

$$^2A_3{}^2T_3 + {}^3A_3{}^3T_3 + {}^4A_3{}^4T_3 = 0. \tag{2.54}$$

Let us also write the boundary conditions for the components of the electric field intensity. For cells 1, 2, and 3 the boundary conditions along axis 3 is

$$a^1E_3 + (1-a)^2E_3 - {}^3E_3 = 0. \tag{2.55}$$

The tangential component of the electric field intensity in cell 3 along axis 3 is equal that of the electric field intensity in cell 4 along axis 3:

$$^3E_3 - {}^4E_3 = 0. \tag{2.56}$$

The normal components of the dielectric displacement of cells 1 and 2 along axis 3 are equal, and hence we can obtain the following expression.

$$^1D_3 - {}^2D_3 = 0.$$

The sum of electric charges on the electrodes of the composite is then equal to the sum of the products of dielectric displacement of the cells over a cross-sectional area

$$^2A_3{}^2D_3 + {}^3A_3{}^3D_3 + {}^4A_3{}^4D_3 = \sigma \,; \tag{2.57}$$

where σ is the charge on the electrodes.

Next, let us find the average values of the strains about directions 1, 2, and 3, which are given by the following formulas:

$$S_1 = a^1S_1 + {}^4S_1 (1-a), \tag{2.58}$$

$$S_2 = {}^4S_2, \tag{2.59}$$

$$S_3 = {}^4S_3. \tag{2.60}$$

As a consequence of then taking into account Eqs. 2.39–2.42, we can obtain

$$({}^1s_{11}{}^1T_1 + {}^1s_{12}{}^1T_2 + {}^1s_{13}{}^1T_3 + {}^1q_{31}{}^1H_3) - (s_{11}{}^2T_1 + {}^2s_{12}{}^2T_2 + {}^2s_{13}{}^2T_3 + {}^2d_{31}{}^2E_3) = 0;$$

$$({}^1s_{21}{}^1T_1 + {}^1s_{22}{}^1T_2 + {}^1s_{23}{}^1T_3 + {}^1q_{32}{}^1H_3) - ({}^2s_{21}{}^2T_1 + {}^2s_{22}{}^2T_2 + {}^2s_{23}{}^2T_3 + {}^2d_{32}{}^2E_3) = 0;$$

$$({}^2s_{11}{}^2T_1 + {}^2s_{12}{}^2T_2 + {}^2s_{13}{}^2T_3 + {}^2d_{31}{}^2E_3) - ({}^3s_{11}{}^3T_1 + {}^3s_{12}{}^3T_2 + {}^3s_{13}{}^3T_3 + {}^3d_{31}{}^3E_3) = 0;$$

$$({}^3s_{31}{}^3T_1 + {}^3s_{32}{}^3T_2 + {}^3s_{33}{}^3T_3 + {}^3d_{33}{}^3E_3) - ({}^4s_{31}{}^4T_1 + {}^4s_{32}{}^4T_2 + {}^4s_{33}{}^4T_3 + {}^4d_{33}{}^4E_3) = 0;$$

$$a({}^1s_{31}{}^1T_1 + {}^1s_{32}{}^1T_2 + {}^1s_{33}{}^1T_3 + {}^1q_{33}{}^1H_3) + (1-a)({}^2s_{31}{}^2T_1 + {}^2s_{32}{}^2T_2$$

$$+ {}^2s_{33}{}^2T_3 + {}^2d_{33}{}^2E_3) - ({}^3s_{31}{}^3T_1 + {}^3s_{32}{}^3T_2 + {}^3s_{33}{}^3T_3 + {}^3d_{33}{}^3E_3) = 0;$$

$$a({}^2s_{21}{}^2T_1 + {}^2s_{22}{}^2T_2 + {}^2s_{23}{}^2T_3 + {}^2d_{32}{}^2E_3) + (1-a)({}^3s_{21}{}^3T_1 + {}^3s_{22}{}^3T_2$$

$$+ {}^3s_{23}{}^3T_3 + {}^3d_{32}{}^3E_3) - ({}^4s_{21}{}^4T_1 + {}^4s_{22}{}^4T_2 + {}^4s_{23}{}^4T_3 + {}^4d_{32}{}^4E_3) = 0;$$

$$(a^2){}^1T_1 + a(1-a){}^2T_1 + (1-a){}^3T_1 = 0;$$

$${}^4T_1 = 0;$$

$$a^3T_2 + (1-a){}^4T_2 = 0;$$

$$a^1T_2 + (1-a){}^2T_2 - {}^3T_2 = 0; \qquad\qquad (2.61)$$

$${}^1T_3 - {}^2T_3 = 0;$$

$$(a^2){}^2T_3 + a(1-a){}^3T_3 + (1-a){}^4T_3 = 0;$$

$$a^1E_3 + (1-a){}^2E_3 - {}^3E_3 = 0;$$

$${}^3E_3 - {}^4E_3 = 0;$$

$$(a^2)({}^2d_{31}{}^2T_1 + {}^2d_{32}{}^2T_2 + {}^2d_{33}{}^2T_3 + {}^2\varepsilon_{33}{}^2E_3) + a(1-a)({}^3d_{31}{}^3T_1 + {}^3d_{32}{}^3T_2$$

$$+ {}^3d_{33}{}^3T_3 + {}^3\varepsilon_{33}{}^3E_3) + (1-a)({}^4d_{31}{}^4T_1 + {}^4d3_2{}^4T_2 + {}^4d_{33}{}^4T_3 + {}^4\varepsilon_{33}{}^4E_3) = \sigma;$$

$$d_{31}E_3 + q_{31}H_3 = S_1; \text{ and}$$

$${}^1q_{31}({}^1T_1 + {}^1T_2) + {}^1q_{33}{}^1T_3 + {}^1\mu_{33}{}^1H_3 = {}^1B_3.$$

The system of equations in 2.61 represents 16 relations that have the same number of unknowns. Since $^4T_I = 0$, the stress can be deleted from all equations of system 2.23; therefore, we are left with 15 equations and the same number of unknowns.

To determine the effective parameters of a composite, it is necessary to add to Eq. 2.61 equations defining the effective parameters of the composite following from Eqs. 2.11 and 2.12:

$$d_{31}E_3 + q_{31}H_3 = S_1; \tag{2.62}$$

$$d_{32}E_3 + q_{32}H_3 = S_2; \tag{2.63}$$

$$d_{33}E_3 + q_{33}H_3 = S_3; \tag{2.64}$$

$$\varepsilon_{33}E_3 + \alpha_{33}H_3 = D_3; \tag{2.65}$$

$$\mu_{33}H_3 + \alpha_{33}E_3 = B_3. \tag{2.66}$$

As a result, we then get a new system of equations, as follows:

$$(^1s_{11}{}^1T_1 + {}^1s_{12}{}^1T_2 + {}^1s_{13}{}^1T_3 + {}^1q_{31}{}^1H_3) - (^2s_{11}{}^2T_1 + {}^2s_{12}{}^2T_2 + {}^2s_{13}{}^2T_3 + {}^2d_{31}{}^2E_3) = 0;$$

$$(^1s_{21}{}^1T_1 + {}^1s_{22}{}^1T_2 + {}^1s_{23}{}^1T_3 + {}^1q_{32}{}^1H_3) - (^2s_{21}{}^2T_1 + {}^2s_{22}{}^2T_2 + {}^2s_{23}{}^2T_3 + {}^2d_{32}{}^2E_3) = 0;$$

$$(^2s_{11}{}^2T_1 + {}^2s_{12}{}^2T_2 + {}^2s_{13}{}^2T_3 + {}^2d_{31}{}^2E_3) - (^3s_{11}{}^3T_1 + {}^3s_{12}{}^3T_2 + {}^3s_{13}{}^3T_3 + {}^3d_{31}{}^3E_3) = 0;$$

$$(^3s_{31}{}^3T_1 + {}^3s_{32}{}^3T_2 + {}^3s_{33}{}^3T_3 + {}^3d_{33}{}^3E_3) - (^4s_{31}{}^4T_1 + {}^4s_{32}{}^4T_2 + {}^4s_{33}{}^4T_3 + {}^4d_{33}{}^4E_3) = 0;$$

$$a(^1s_{31}{}^1T_1 + {}^1s_{32}{}^1T_2 + {}^1s_{33}{}^1T_3 + {}^1q_{33}{}^1H_3) + (1-a)(^2s_{31}{}^2T_1 + {}^2s_{32}{}^2T_2$$
$$+ {}^2s_{33}{}^2T_3 + {}^2d_{33}{}^2E_3) - (^3s_{31}{}^3T_1 + {}^3s_{32}{}^3T_2 + {}^3s_{33}{}^3T_3 + {}^3d_{33}{}^3E_3) = 0;$$

$$a(^2s_{21}{}^2T_1 + {}^2s_{22}{}^2T_2 + {}^2s_{23}{}^2T_3 + {}^2d_{32}{}^2E_3) + (1-a)(^3s_{21}{}^3T_1 + {}^3s_{22}{}^3T_2$$
$$+ {}^3s_{23}{}^3T_3 + {}^3d_{32}{}^3E_3) - (^4s_{21}{}^4T_1 + {}^4s_{22}{}^4T_2 + {}^4s_{23}{}^4T_3 + {}^4d_{32}{}^4E_3) = 0;$$

$$(a^2)\,{}^1T_1 + a(1-a)\,{}^2T_1 + (1-a)\,{}^3T_1 = 0;$$

$$a^3T_2 + (1-a)^4\,{}^4T_2 = 0;$$

$$a^1T_2 + (1-a)^2\,{}^2T_2 - {}^3T_2 = 0. \tag{2.67}$$

$$^1T_3 - {}^2T_3 = 0;$$

$$(a^2)\,{}^2T_3 + a(1-a)^3T_3 + (1-a)^4T_3 = 0;$$

$$a^1E_3 + (1-a)^2E_3 - {}^3E_3 = 0;$$

$$^3E_3 - {}^4E_3 = 0;$$

$$(a^2)({}^2d_{31}{}^2T_1 + {}^2d_{32}{}^2T_2 + {}^2d_{33}{}^2T_3 + {}^2\varepsilon_{33}{}^2E_3) + a(1-a)T^3d_{31}{}^3T_1 + {}^3d_{32}{}^3T_2$$
$$+ {}^3d_{33}{}^3T_3 + {}^3\varepsilon_{33}{}^3E_3) + (1-a)({}^4d_{31}{}^4T_1 + {}^4d3_2{}^4T_2 + {}^4d_{33}{}^4T_3 + {}^4\varepsilon_{33}{}^4E_3) = \sigma;$$

$$d_{31}E_3 + q_{31}H_3 = S_1;$$

$$d_{32}E_3 + q_{32}H_3 = S_2;$$

$$d_{33}\,E_3 + q_{33}H_3 = S_3;$$

$$\varepsilon_{33}E_3 + \alpha_{33}H_3 = D_3;$$

$$\mu_{33}H_3 + \alpha_{33}E_3 = B_3; \quad \text{and}$$

$$^1q_{31}({}^1T_1 + {}^1T_2) + {}^1q_{33}{}^1T_3 + {}^1\mu_{33}{}^1H_3 = {}^1B_3.$$

In addition, the average values of the strains can be obtained by the following formulas:

$$S_1 = a({}^1s_{11}{}^1T_1 + {}^1s_{12}{}^1T_2 + {}^1s_{13}{}^1T_3 + {}^1q_{31}{}^1H_3)$$
$$+ ({}^4s_{12}{}^4T_2 + {}^4s_{13}{}^4T_3 + {}^4d_{31}{}^4E_3)(1-a), \tag{2.68}$$

$$S_2 = {}^4s_{22}{}^4T_2 + {}^4s_{23}{}^4T_3 + {}^4d_{32}{}^4E_3, \tag{2.69}$$

$$S_3 = {}^4s_{32}{}^4T_2 + {}^4s_{33}{}^4T_3 + {}^4d_{33}{}^4E_3. \tag{2.70}$$

The system of equations in 2.67, in view of 2.68–2.70, can readily be solved using available computational software packages, such as Maple 7. The material parameters presented earlier in Table 2.4 has sufficient information to perform these calculations for 0–3 composites. To determine the effective parameters of the solution to the system of equations in 2.67, the algorithm presented in Fig. 2.17 can be used.

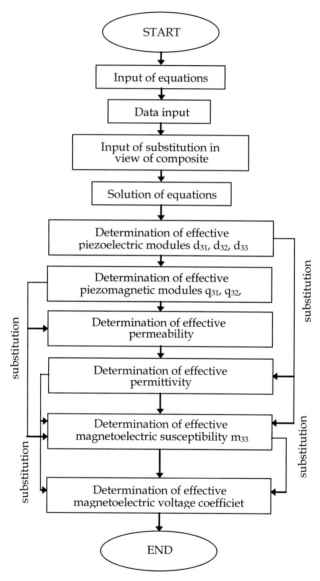

Figure 2.17 Algorithm for determination of composite effective parameters.

To account for the external forces that act on the cells in the cubic model, it is necessary to consider the clamped condition. Thus, to the components of the stress tensor in the system of Eq. 2.67, one must add additional external stress components. In addition, it is necessary to use boundary conditions of the type

$$S_1 = -s_{c1}\, s_{11}\, T_1,$$
$$S_2 = -s_{c2}\, s_{22}\, T_2, \qquad\qquad (2.71)$$
$$S_3 = -s_{c3}\, s_{33}\, T_3,$$

where $s_{ci} = {}^c s_{ii}/s_{ii}$ is the relative compliance of the clamp, and ${}^c s_{ii}$ is compliance of the clamp.

Because of inconveniences in the analytical expressions for effective parameters of composites, computer calculations of the dependence of effective parameters on the relative piezoelectric phase volume in ME composite have been performed. Consider that the volume of a cubic cell defined as the sum of volumes magnetostrictive and piezoelectric phases is equal 1, that the volume of the magnetostrictive phase is equal to a^3, and that the volume of the piezoelectric phase is equal to $(1 - a^3)$. In this case, the volume fraction of the piezoelectric phase in the ME composite is as follows:

$$v = V_P/(V_P + V_M) = (1 - a^3)/1 = 1 - a^3, \qquad\qquad (2.72)$$

where V_P is the volume of the piezoelectric phase, and V_M that of the magnetostrictive phase.

To determine the effective piezoelectric modules, suppose that the composite is under an external electric field, and that the external magnetic field is absent (i.e. $H_3 = 0$). Then, to solve Eq. 2.67, let us consider that the average values of electric and magnetic fields in a cubic cell are defined by

$$E_3 = {}^4E_3, \qquad\qquad (2.73)$$
$$H_3 = {}^4H_3.$$

The resultant outcomes for an evaluation of effective piezoelectric modules are then given in Figs. 2.18 and 2.19. From these figures, it follows in the limiting cases (i.e., $v = 0$ and $v = 1$) that the effective piezoelectric modules coincides with that of the composite components which were given in Table 2.4.

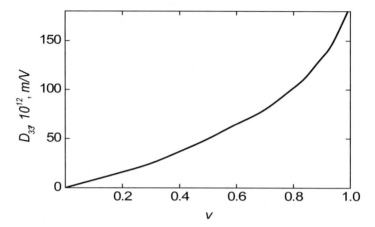

Figure 2.18 Concentration dependence of the piezoelectric module d_{33}.

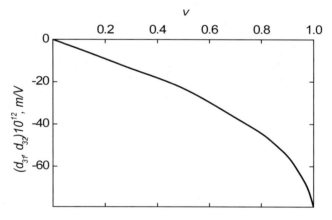

Figure 2.19 Concentration dependence of the piezoelectric module d_{31}.

To determine the effective piezomagnetic modules, let us suppose that the composite is under an external magnetic field, and that an external electric field is absent (i.e., $E_3 = 0$). In this case, the solution to Eq. 2.76 allows us to obtain the dependence of the effective piezomagnetic modules on the piezoelectric phase volume fraction, as given in Figs. 2.20 and 2.21. In these figures, the effective piezomagnetic modules in limiting cases (i.e., $v = 0$ and $v = 1$) can be seen to coincide with those of initial composite components which were given earlier in Table 2.4.

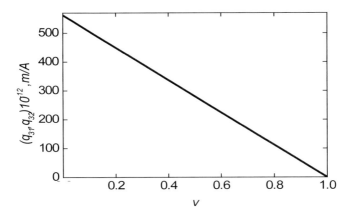

Figure 2.20 Concentration dependence of effective piezomagnetic modules q_{31} *and* q_{32}.

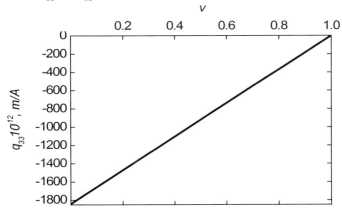

Figure 2.21 Concentration dependence of effective piezomagnetic module q_{33}.

To determine the effective magnetic permeability, let us suppose that $E_3 = 0$ and that the composite has a magnetic induction of

$$B_3 = {}^1B_3 a^2 + {}^3H_3 \mu_0 a(1-a) + {}^4H_3 \mu_0 (1-a). \qquad (2.74)$$

Calculations following Eq. 2.67 leads to the dependence of the effective permeability on piezoelectric phase volume fraction as shown in Fig. 2.22.

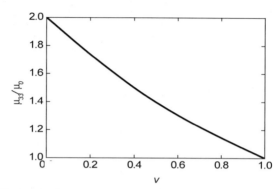

Figure 2.22 Graph of concentration dependence μ_{33} on v.

To determine the effective dielectric permittivity, let us suppose that $H_3 = 0$ and that the composite has an electric induction of $D_3 = \sigma$. Considering the value for the effective piezoelectric modulus obtained earlier, we can get the dependence of the effective permittivity on the piezoelectric phase volume fraction, as shown in Fig. 2.23. In this figure, the limiting case ($v = 0$ and $v = 1$) values of the effective permittivity coincide with that of the composite components presented earlier in Tables 2.3 and 2.4.

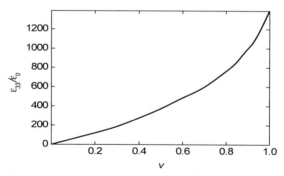

Figure 2.23 Concentration dependence of an effective permittivity.

To determine the effective ME susceptibility, supposed that the composite has a midvalue of the magnetic induction equal to zero, and that it has a midvalue of the electric induction equal to $D_3 = \sigma$. For this case, the solution to Eq. 2.67 allows us to find the dependence of the effective ME susceptibility on the piezoelectric phase volume fraction, which is shown in Fig. 2.24.

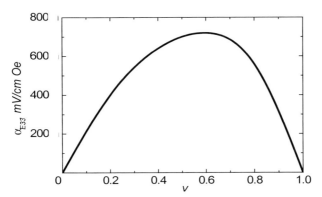

Figure 2.24 Concentration dependence of ME susceptibility for composite with connectivity 3-0.

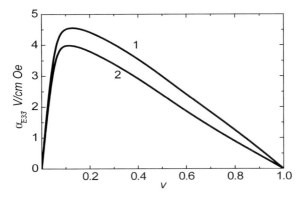

Figure 2.25 Graph of concentration dependence of ME voltage coefficient for composite with connectivity 3–0: 1 — calculation on the offered model, 2 — calculation on model [7] for $\alpha_{E33} = E_3/{}^mH_3$.

The dependence of the effective ME voltage coefficient, defined as

$$\alpha_{E,33} = -\alpha_{33}/\varepsilon_{33}, \tag{2.75}$$

on the piezoelectric phase volume fraction can then be easily obtained, as shown in Fig. 2.25. These graphical solutions then allow us to determine the piezoelectric and magnetostrictive phase volume fractions that yield maximum values for the effective ME susceptibility (Fig. 2.24) and the ME voltage coefficient (Fig. 2.25).

Calculations of $\alpha_{E,33} = E_3/{}^m H_3$ have also been performed for electric and magnetic fields applied for free composites using the material parameters in [7] (see Fig. 2.26). The values of the ME voltage coefficient in Fig. 2.26 coincide with previously published data [7], demonstrating the usefulness of the predictions. As follows from Fig. 2.25, the ME voltage coefficient was approximately 20% greater than that calculated from the experimental data using the model. This is explained by the fact that the internal (local) magnetic field in the ferrite component is considerably different than that of the externally applied magnetic field.

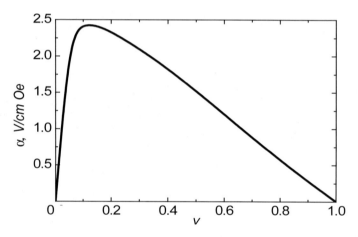

Figure 2.26 Graph of concentration dependence $\alpha_{E33} = E_3/{}^m H_3$ for composite with material parameters from [7].

2.2.1.2 Composite with connectivity 0–3

The effective parameters of a composite with a connectivity 0–3 were found by analogy to the previous sections. Figures 2.27 and 2.28 show the dependence of the ME susceptibility and the ME voltage coefficient on the piezoelectric phase volume fraction. The results for the 0–3 and 3–0 composites were similar. This is because the connectivity of 0–3 differs from that of 3–0 only by commutation of the piezoelectric and magnetostrictive phases (see Fig. 2.15).

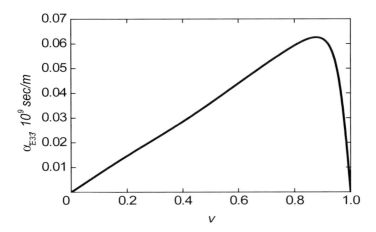

Figure 2.27 Concentration dependence of ME voltage coefficient for composite with connectivity 0–3.

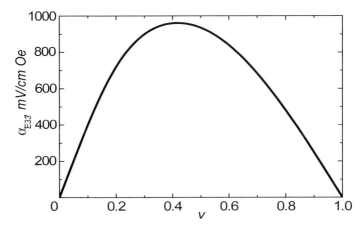

Figure 2.28 Concentration dependence of ME susceptibility of composites with connectivity 0–3.

2.2.1.3 ME effect in clamped samples

The ME parameters of the models considered above are fulfilled for mechanically free grains of the composite components. However, in real ferrite–piezoelectric ceramic composites, the component phase grains are mechanically clamped by the neighboring grains and by environmental conditions. In this case, mechanical clamping can be accounted for in Eqs. 2.62–2.66 by including nonzero midvalues of the stress tensor T_i (i = 1, 2, 3). Thus, $S_i = s_{cii}T_i$, where $s_{cii} = S_i/T_i$ is the compliance of the equivalent system. For free composites, $s_{cii} \gg s_{ii}$; and for clamped ones $s_{cii} = 0$.

Calculations using Eq. 2.67, in view of comments just above, have shown that the magnitude of the ME susceptibility and the ME voltage coefficient strongly depend on the degree of clamping in the component grains. For a totally clamped composite, these two effective parameters have been shown to be decreased by nearly two orders of magnitude (Fig. 2.29). The decrease of the ME parameters with increasing degree of clamping is due to partial blocking of the piezoelectric and piezomagnetic effects.

2.2.2 Experimental Data

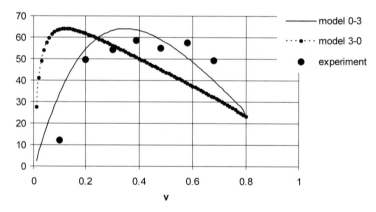

Figure 2.29 ME effect in bulk composite NFO–PZT.

Measurements of the ME voltage coefficient have been performed for bulk composites of NFO–PZT, using the experimental methodology mentioned above. Data are shown in Fig. 2.29 for the ME voltage coefficient as a function of the piezoelectric phase volume fraction. These measured values are much lower in magnitude than the theoretical ones predicted for the free composite condition. However, considering a clamped condition defined by the matching $s_{c11} = s_{c22} = s_{c33} = 0.3s_{33}$, agreement between theory and experiment was found, as illustrated in Fig. 2.29. These results indicate in real 0–3 ferrite–piezoelectric ceramic composites mixtures that the component phase grains are mechanically clamped by neighboring grains and by environmental boundary conditions.

2.3 MAXWELL–WAGNER RELAXATION IN ME COMPOSITES

Ferrite–piezoelectric composites can be considered as a thermodynamic system that are capable of responding to an infinitesimal externally applied field. Relaxation can then be considered as the self-adjustment of the thermodynamic system in response to an external force [22] that maintains equilibrium. It reflects an adjustment of internal parameters to the balanced values. Specifically, dielectric relaxation is a time-dependent induced polarization P response to an applied electric field E [23, 24]. The process is characterized by dispersion in the internal parameter $\delta P/\delta E$. Polar dielectrics with low resistivities often have a strongly enhanced permittivity at low frequencies, due to charge separation. Such space charge polarization contributions to the permittivity are characterized by low-frequency dispersion [25]. Investigations have shown that the effects of space charge polarization relaxation reveals itself not only in the dielectric permittivity, but also in the elastic constants of layered polar dielectrics [25].

In the case of dielectric relaxation in the quasistatic frequency range, when the applied electric field duration period may exceed times of seconds or more, space charge polarization contributions to the dielectric permittivity have sufficient time to respond to the drive: consequently, the permittivity has a constant value. However, as the frequency is increased, this polarization mechanism can not respond efficiently to the drive, rather there is a phase angle between drive and response. The tangent of this phase angle is the dielectric loss factor. Thus, the material is characterized by the complex dielectric permittivity that depends on frequency

$$\varepsilon(\omega) = \varepsilon'(\omega) - i\varepsilon''(\omega), \tag{2.76}$$

where $\varepsilon'(\omega)$ is the real part (i.e., the permittivity of storage compliance), and $\varepsilon''(\omega)$ is imaginary part (i.e., the phase angle or energy loss). Dynamic experiments in which *the electric field* applied to the material is periodically varied with a circular frequency are standard methods for measuring the dielectric constant. There are many impedance analyzers capable of measuring the complex frequency-dependent dielectric constant over frequencies extending from micro to mega Hertz.

Since dynamic response functions allow for the description of frequency-dependent materials parameters, they are suitable for use in the analysis of the ME susceptibility and voltage coefficient of ferrite–piezoelectric composites, both of which are complex functions of frequency. Investigations of the relaxation phenomenon of ferrite–piezoelectric composites have previously been reported.

The relations that we obtained for the effective parameters of composite in the last section did not consider their frequency dependence. Calculations were done using the ferroelectric polarization, which will provide a fair estimate of the high-frequency polarization. Accumulation of electric charges at the boundaries of composite components will result in additional contributions to the polarization: albeit low-frequency ones. Said accumulation of free charges at the interfaces between component phases will result in dielectric dispersion and losses under low-frequency alternating fields, which is known as Maxwell-Wagner relaxation. The purpose of the following section is an analysis of Maxwell-Wagner relaxation of the effective parameters in ferrite–piezoelectric composites, in particular, the ME susceptibility and voltage coefficient.

2.3.1 Layered Composites

Now, let us consider Maxwell-Wagner relaxation of the ME parameters for a multilayer composite with a 2–2 type connectivity [26, 27]. Following from Eq. 2.76, the specific equations, in view of finite electrical conductivities of component phases, for the complex permittivity of the components is

$$^P\varepsilon_{33} = {}^P\varepsilon - i^P\gamma/\omega,$$

$$^m\varepsilon_{33} = {}^m\varepsilon - i^m\gamma/\omega; \tag{2.77}$$

where $^P\gamma$ and $^m\gamma$ are the conductivities of the piezoelectric and magnetostrictive phases; and ω the circular frequency, where the frequency is less than that of electromechanical resonance. The effective parameters of the composite can be then determined by an average of expressions for the components of strain, electric, and magnetic inductions, following Eqs. 2.7–2.10. Furthermore, assume that the component phase layers are thin and arranged in the plane OX_1X_2; that the piezoelectric component is polarized along the OX_3 axis, along which same axis the an electric field with a circular frequency ω is applied; and that a magnetic bias and variable magnetic fields are applied along the OX_1 axis. In this case, the boundary conditions are similar to those given in Eq. 2.14.

The general characteristics of the frequency-dependent ME susceptibility will fulfill the Debye formula [23]:

$$\alpha_{13} = \alpha'_{13} - i\alpha''_{13}, \tag{2.78}$$

$$\alpha'_{13} = \alpha_{130} + \Delta\alpha_{13}/(1 + \omega^2\tau_a^2); \quad \alpha''_{13} = \Delta\alpha_{13}\,\omega\tau_a/(1 + \omega^2\tau_a^2),$$

where $\Delta\alpha_{13} = \alpha_{130} - \alpha_{13\infty}$ is the relaxation strength, α_{130} and $\alpha_{13\infty}$ are the static ($\omega \to 0$) and high-frequency ($\omega \to \infty$) ME susceptibilities, and τ_a the relaxation time. The transverse static and high-frequency ME susceptibilities, and in addition the relaxation time, can be found from the solutions of Eqs. 2.7–2.13 by taking into account Eqs. 2.14 and 2.77. Then, by supposing that the symmetry of the piezoelectric phase is ∞m, and that the magnetic phase possesses cubic symmetry, we obtain the following expressions:

$$\alpha_{130} = \frac{^m\gamma vk(1-v)(^mq_{12} + {}^mq_{11})^Pd_{31}}{\left[[^P\gamma(1-v) + {}^m\gamma v][(^ms_{12} + {}^ms_{11})v + (^Ps_{11} + {}^Ps_{12})(1-v)]\right]}, \tag{2.79}$$

$$\alpha_{13\infty} = \frac{^m\varepsilon vk(1-v)(^mq_{12} + {}^mq_{11})}{\left[[v{}^m\varepsilon + {}^P\varepsilon(1-v)][(^ms_{12} + {}^ms_{11})v + (^Ps_{11} + {}^Ps_{12})(1-v)] - 2^Pd_{31}^2(1-v)^2\right]},$$

$$\tau_\alpha = \frac{{}^m\varepsilon v + {}^p\varepsilon(1-v)}{{}^m\gamma v + {}^p\gamma(1-v)} - \frac{2k(1-v)^2\,{}^pd_{31}^2}{\left[v\left({}^ms_{11}+{}^ms_{12}\right)+\left({}^ps_{11}+{}^ps_{12}\right)(1-v)\right]\left[{}^m\gamma v + {}^p\gamma(1-v)\right]}.$$

As an example, we shall consider the composite to consist of polarized ferroelectric PZT ceramics and nickel ferro-spinel. In numerical calculations, the following values of the materials parameters for composite phase components can be used:

${}^ps_{11} = 15.3\cdot10^{-12}\,\text{m}^2/\text{N}$, ${}^ps_{12} = -5\cdot10^{-12}\,\text{m}^2/\text{N}$, ${}^ps_{13} = -7.22\cdot10^{-12}\,\text{m}^2/\text{N}$,

${}^ps_{33} = 17.3\cdot10^{-12}\,\text{m}^2/\text{N}$, ${}^ms_{11} = 15.3\cdot10^{-12}\,\text{m}^2/\text{N}$, ${}^ms_{12} = -5\cdot10^{-12}\,\text{m}^2/\text{N}$,

${}^mq_{33} = -1880\cdot10^{-12}\,\text{m}/\text{A}$, ${}^mq_{31} = 556\cdot10^{-12}\,\text{m}/\text{A}$, ${}^pd_{31} = -175\cdot10^{-12}\,\text{m}/\text{V}$,

${}^pd_{33} = -400\cdot10^{-12}\,\text{m}/\text{V}$, ${}^m\mu_{33}/\mu_0 = 3$, ${}^p\varepsilon/\varepsilon_0 = 1750$, ${}^m\varepsilon/\varepsilon_0 = 10$,

${}^m\gamma = 10^{-5}\,(\text{Ohm}\cdot\text{m})^{-1}$, ${}^p\gamma = 10^{-13}\,(\text{Ohm}\cdot\text{m})^{-1}$.

Relaxation of the effective permittivity of the composite was then calculated, and is shown in Fig. 2.30. Under the conditions of ${}^p\gamma/{}^m\gamma \ll 1$, ${}^p\varepsilon/{}^m\varepsilon \gg 1$, and $v \ll 1$, it is possible to obtain a giant increase in the static dielectric permittivity. The results are in agreement with prior reports for ferroelectric–polymer composites [24]. In the work given in Fig. 2.30, the dependence of the parameters for dielectric relaxation depended on an interface connection parameter. The strength of the relaxation was found to decrease with increase of this interphase–interface parameter.

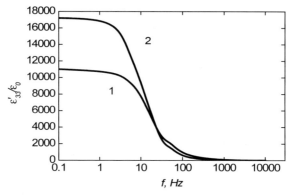

Figure 2.30 Dielectric relaxation in composite for v = 0.1: 1 — k = 1, 2 — k = 0.

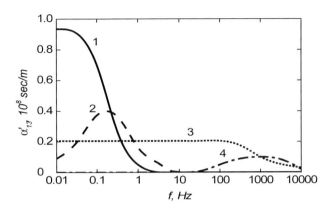

Figure 2.31 Frequency dependence real (1, 3) and imaginary (2, 4) parts of a ME susceptibility: 1, 2 — $v = 0.001$, 3, 4 — $v = 0.9$.

Figure 2.31 illustrates the dependence of the relaxation of the ME susceptibility on piezoelectric phase volume fraction depth. A large relaxation strength is characteristic for a composite whose piezoelectric phase component has a big permittivity and whose ferrite one is electrically conductive. In the case of $^m\varepsilon/^p\varepsilon \ll 1$ and $^p\gamma/^m\gamma \ll 1$, a maximum value of the relaxation strength is observed at a piezoelectric phase volume fraction of $v_1 \approx [(^ps_{11} + {}^ps_{12})/({}^ms_{11} + {}^ms_{12})]^{1/2} (^p\gamma/^m\gamma)^{1/2}$. If the compliances of both composite phase components are taken as equal, then one gets $v_1 \approx (^p\gamma/^m\gamma)^{1/2}$.

Relaxation of the ME susceptibility is shown in Fig. 2.32. For $^p\gamma/^m\gamma \ll v \ll 1$, the static ME susceptibility approaches a value equal to $(^mq_{12} + {}^mq_{11})\ ^pd_{31}/(^ps_{11} + {}^ps_{12})$. For a composite with a composition dictated by the condition $v_1 = (^ph/^mh)^{1/2} \approx 10^{-4}$ where ^ph and ^mh are the thicknesses of the piezoelectric and ferrite phase components, the value of the ME susceptibility is equal $0.94 \cdot 10^{-8}$ s/m. This large magnitude of the static ME susceptibility is due to a large local electric field in the thin piezoelectric phase component, to the significant electrical conductance of the ferrite layer, and to the large internal mechanical stresses ^pT_j and ^mT_j ($j = 1,2$) that are induced by the electrical field in the piezoelectric component. The predicted maximum value for the ME susceptibility of this composite exceeds the experimentally observed one for known materials parameters.

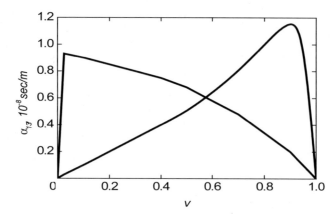

Figure 2.32 Concentration dependence of static and high-frequency magnetoelectric susceptibilities.

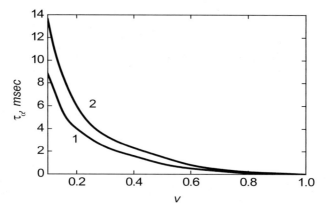

Figure 2.33 Concentration dependence of relaxation time of ME susceptibility: 1 — $k = 1$, 2 — $k = 0.1$.

In the case of a weak piezoelectric effect considered here as $\left(^{p}d^{2}_{31}/[(^{p}s_{11} + {}^{p}s_{12})\,{}^{p}\varepsilon]\right) << 1$), the relaxation time as follows from Eq. 2.79 at $^{p}\gamma/^{m}\gamma << v/(1 - v)$ is controlled mainly by the charging time of the piezoelectric layer capacitance, across the resistance of the ferrite layer: $\tau_{\alpha} \approx (^{p}\varepsilon/^{m}\gamma)(1 - v)/v$, as shown in Fig. 2.33. With increase of the piezoelectric phase volume fraction, the relaxation time decreases rapidly, whereas with increase of the piezoelectric constant of the ferroelectric phase, that of the composite also increases. The stresses ${}^{p}T_{1}$ and ${}^{p}T_{2}$ (caused by a transverse piezoelectric effect) induce additional charge on the piezoelectric layer, increase the charging time of its capacity, and decrease the relaxation time of the ME susceptibility.

Next, we consider the effect that the interfacial parameter k has on the relaxation time. As follows from Eq. 2.79, the relaxation time τ_{α} and relaxation frequency $\omega_{r} = 1/\tau_{\alpha}$ (determined by the frequency at which maximum in the imaginary ME susceptibility occurs) can be predicted for a wide range of volume fractions of the component phases, and also for various choices of materials for the component phases. Since the frequency dependence of the transverse ME voltage coefficient is defined by the Debye formulas, we can arrive at relations analogously to Eq. 2.79 which provides us the dependence of the relaxation time on various parameters, as follows:

$$\alpha_{E,T} = \alpha'_{E,T} - i\alpha''_{E,T},$$
$$\alpha'_{E,T} = \alpha_{E,T\infty} + \Delta\alpha_{E,T}/(1 + \omega^2 \tau^2_{\alpha T}); \qquad (2.80)$$
$$\alpha''_{E,T} = \Delta\alpha_{E,T}\,\omega\,\tau_{\alpha T}/(1 + \omega^2 \tau^2_{\alpha T});$$

where $\Delta\alpha_{E,T} = -\alpha_{E,T\infty}$

$$= \frac{v^{p}d_{31}k(1-v)(^{m}q_{12} + {}^{m}q_{11})}{\left[(^{m}s_{12} + {}^{m}s_{11})v + (^{p}s_{12} + {}^{p}s_{11})(1-v)\right] - 2^{p}d^2_{31}(1-v)},$$

$$\tau_{\alpha T} = \frac{{}^{p}\varepsilon}{{}^{p}\gamma} - \frac{2k(1-v)^{p}d^2_{31}}{{}^{p}\gamma[v(^{m}s_{11} + {}^{p}s_{m12}) + (^{p}s_{11} + {}^{p}s_{11})(1-v)]}.$$

Following this relationship, we can see in Fig. 2.33 that increasing k has the same effect on relaxation time as previously shown for increasing piezoelectric constant: it increases notably the value of τ_{α}.

Unlike the ME susceptibility, the real part of the ME voltage coefficient increases with increasing frequency: i.e., inverse relaxation occurs, as shown in Fig. 2.34. The relaxation strength was maximum for $v \approx 0.5$. In the case of a weak piezoelectric effect (defined as $d^2_{31}/[(^{p}s_{11} + {}^{p}s_{12})^{p}\varepsilon] \ll 1$), the relaxation time is defined controlled by the discharge time of the piezoelectric layer capacitance through its own resistance $\tau_{\alpha T} \approx {}^P\varepsilon/{}^P\gamma$, and does not depend on volume fraction of the component phases. In this case, it can be shown that $\tau_{\alpha T} >> \tau_\alpha$, as illustrated in Fig. 2.35. This figure also shows that increasing piezoelectric phase volume fraction decreases $\tau_{\alpha T}$. This diminution is due to the appearance of additional charge on the piezoelectric layer capacitance under effect of stresses ${}^P T_1$ and ${}^P T_2$, which is induced by external magnetic fields applied to the composite.

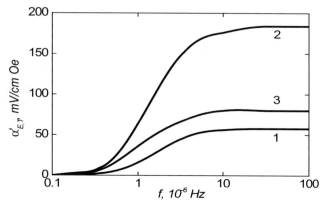

Figure 2.34 Frequency dependence of real part of ME voltage coefficient: $1 — v = 0.1, 2 — v = 0.5, 3 — v = 0.9$.

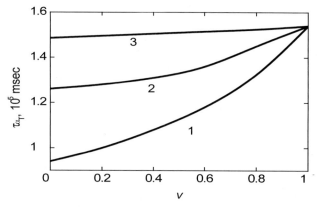

Figure 2.35 Concentration dependence of relaxation time of ME voltage coefficient: $1 — k = 1, 2 — k = 0.5, 3 — k = 0.1$.

Thus, in layered ferrite–piezoelectric composites, a strong relaxation of the ME susceptibility is observed, whereas, for the ME voltage coefficient, a strong inverse effect occurs. The relaxation time and relaxation frequency of the ME susceptibility can be changed over a wide range of values, by varying the volume fractions of the composite's component phases, and also by selection of the initial component phases as different piezoelectric and magnetostrictive phases can have different values of property tensors.

2.3.2 Bulk Composites

As an example, consider the cubic model of bulk ferrite–piezoelectric composites [7] in which one phase has connectivity in all three directions and in which a second isolated phase does not have connectivity in any direction. Following known classifications [1], this specified composite has a connectivity of the type 3–0. For the piezoelectric and magnetostrictive phases, the equations for the strain, dielectric displacement, and the magnetic induction were presented earlier in Eqs. 2.39–2.42. In addition, in view of a finite electrical conductivity, the permittivities of the component phases are given by Eq. 2.77.

The general formulas for defining the effective parameters of the composite can then be obtained by averaging the expressions for components of the strains, and the electric and magnetic inductions following Eqs. 2.58–2.60. Then assume that the sample has the form of a disk that is oriented in the plane OX_1X_2, where the piezoelectric phase is polarized along the axis OX_3, an electric field with a frequency ω is applied along the same axis, and both a magnetic bias and an alternate magnetic field are applied along axis OX_1.

Static and high-frequency ME susceptibilities, in addition to the relaxation time, can be obtained from solutions to Eq. 2.77 by applying boundary conditions for strains, stresses, and electric and magnetic fields. Then, suppose that the symmetry of the piezoelectric phase is ∞m, and that the magnetic phase possesses cubic symmetry. An analytical expression can then be obtained, but due to inconvenience of its form, solution of equations can only be fulfilled numerically [31, 32]. As an example system to apply this numerical solution, we shall consider a composite consisting of poled PZT ferroelectric ceramics and nickel ferro-spinel, which has the following values of comphonent phase material parameters:

$^{p}s_{11} = 15.3\cdot10^{-12}\ m^2/N$, $^{p}s_{12} = -5\cdot10^{-12}\ m^2/N$, $^{p}s_{13} = -7.22\cdot10^{-12}\ m^2/N$,

$^{p}s_{33} = 17.3\cdot10^{-12}\ m^2/N$, $^{m}s_{11} = 15.3\cdot10^{-12}\ m^2/N$, $^{m}s_{12} = -5\cdot10^{-12}\ m^2/N$,

$^{m}q_{33} = -1880\cdot10^{-12}\ m/A$, $^{m}q_{31} = 556\cdot10^{-12}\ m/A$, $^{p}d_{31} = -50\cdot10^{-12}\ m/V$,

$^{p}d_{33} = -100\cdot10^{-12}\ m/V$, $^{m}\mu_{33}/\mu_0 = 2$, $^{p}\varepsilon/\varepsilon_0 = 1000$, $^{m}\varepsilon/\varepsilon_0 = 10$, $^{m}\gamma = 10^{-5}$

$(Ohm\cdot m)^{-1}$, $^{p}\gamma = 10^{-13}\ (Ohm\cdot m)^{-1}$.

In Figs. 2.36 and 2.37, the frequency dependences of the effective piezoelectric modules of this 0-3 composite is illustrated. With increase of the piezoelectric phase volume fraction, the relaxation strength increases, and the relaxation frequency decreases. Thus, parameters of the piezoelectric relaxation can be varied by changing the relative volume fractions of the composite component phases.

The relaxation of the effective dielectric permittivity is illustrated in Fig. 2.38. Under the conditions of $^{p}\gamma/^{m}\gamma << 1$, $^{p}\varepsilon/^{m}\varepsilon >> 1$, and $v << 1$, pronounced increases in the static permittivity can be seen, similar to that found earlier for layered composites. In this case, the relaxation frequency does not depend much on the volume fraction of the components, and is mainly defined by electric parameters. Interestingly, as shown in Fig. 2.39, there is a frequency for which the effective permittivity does not depend much on the piezoelectric phase volume fraction.

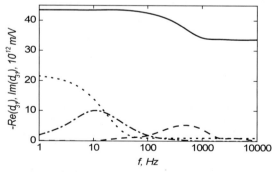

Figure 2.36 Frequency dependence of real (1, 3) and imaginary (2, 4) parts of the effective piezoelectric module d_{31}: 1, 2 — $v = 0.9$, 3, 4 — $v = 0.1$.

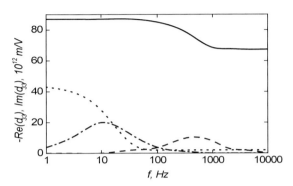

Figure 2.37 Frequency dependence of real $(1, 3)$ and imaginary $(2, 4)$ parts of the effective piezoelectric module d_{33}: $1, 2 — \upsilon = 0.9, 3, 4 — \upsilon = 0.1$.

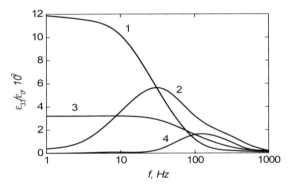

Figure 2.38 Frequency dependence of real $(1, 3)$ and imaginary $(2, 4)$ parts of effective permittivity: $1, 2 — \upsilon = 0.1, 3, 4 — \upsilon = 0.9$.

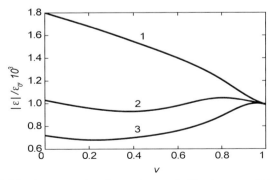

Figure 2.39 Concentration dependence of effective permittivity module: $1 — f = 200$ Hz, $2 — f = 350$ Hz, $3 — f = 500$ Hz.

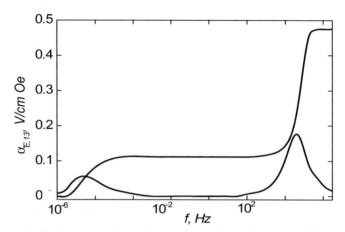

Figure 2.40 Frequency dependence of real (1) and imaginary (2) parts of ME voltage coefficient for $v = 0.6$.

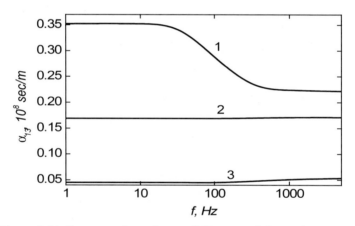

Figure 2.41. Frequency dependence of ME susceptibility real part: 1 — $v = 0.5$, 2 — $v = 0.8$, 3 — $v = 0.95$.

The frequency dependence of the ME voltage coefficient is shown in Fig. 2.40. The real part of the ME voltage coefficient can be seen to increase with increasing frequency, i.e., an inverse relaxation occurs. The relaxation spectrum is characterized by two regions of dispersion. The first cover the range of infra-low frequency. The relaxation time for this region depends on weakly on the component phase volume fractions, and is controlled by the discharge time of the piezoelectric phase capacitance (see Fig. 2.16) through its own resistance. The second region of relaxation covers the frequency range over which the voltages on capacitance elements 2, 3, and 4 are equalized. In this region, the relaxation is defined by the charging time of the capacitance of element 2 through the resistance of the ferrite phase component, and it varies as the relative volume fraction of the phases is changed.

Relaxation of the ME susceptibility is then shown in Fig. 2.41. The relaxation time of the ME susceptibility is defined by the charging time of the capacitance of element 2, through the resistance of the ferrite component. The relaxation strength decreases with increase of the piezoelectric component volume fraction, and for volumes of $v > 0.8$ becomes negative. Thus, relaxation of the ME susceptibility depends on the volume fraction of component phases, and can be either direct or inverse. Interestingly, there is a specific value for the piezoelectric phase volume fraction for which there is no frequency dependence of the ME susceptibility. This specific value is defined by composite component parameters, as illustrated in Fig. 2.41. The static and high-frequency ME susceptibilities are then shown in Fig. 2.42. It can be seen that the static ME susceptibility has a peak for $v \ll 1$.

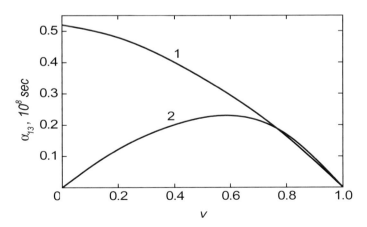

Figure 2.42 Dependence of static (1) and high-frequency (2) ME susceptibilities on volume fraction of the piezoelectric component.

2.4 CONCLUSIONS

In this chapter, a generalized theoretical model for low-frequency ME effects in layered composites was introduced. To describe the composite's physical properties, effective parameters were derived. This model allowed us to define on the basis of an exact solution the effective mechanical, electric, magnetic, and ME parameters of layered ferrite–piezoelectric composite. Expressions for the effective parameters, (including ME susceptibility and ME voltage coefficient) were derived as functions of an interface coupling parameter, constituent phase material parameters, and relative volume fractions of phases. Longitudinal, transverse and in-plane cases were all considered.

The approach predicted giant ME effects in ferrite–PZT ceramic composites. Predictions of the ME effect for various model ceramic composite systems were given including CFO–BTO, CFO–PZT, NFO–PZT, and lanthanum strontium manganite–PZT. It was shown that ME effect in ferrite–PZT systems is maximum for magnetic and electric fields applied along an in-plane orientation: however, this has yet to be experimentally verified. Using the model, a comparison of the ME parameters was then made between calculated values and experimental data. From this comparison, the importance of an interfacial coupling parameter between phases was inferred. This interphase interfacial connection parameter was shown to be weak for CFO–PZT and lanthanum strontium manganite–PZT composites, but near ideal for NFO–PZT.

In addition, the generalized theory allows for modeling of the low-frequency ME effect in bulk composites. To describe these low-frequency composite properties, an effective medium method was used. Calculation of the effective permittivity and permeability, piezoelectric and piezomagnetic modules, ME susceptibility, and ME voltage coefficient were performed as functions of volume fractions and component parameters. Composites with dimensional connectivities of types 3–0 and 0–3 were considered.

Larger ME coefficients were found for 3–0 composites with magnetic and/or electric fields applied along the longitudinal direction. For composites of CFO–PZT, values as high as 4 V/(cmOe) were predicted for the ME voltage coefficient. Fields applied along transverse orientations were found to have ME effects about 2–3.5× smaller than those for longitudinally applied fields. Furthermore, clamping was shown to significantly reduce the ME effect. Finally, in bulk ferrite–piezoelectric composites, large relaxation effects in the ME susceptibility were found with increasing frequency. The relaxation time and frequency of the ME susceptibility and voltage coefficient can be varied by changing the volume fraction of component phases or by changes in the composite component parameters.

Acknowledgments

This work was supported by the Russian Foundation for Basic Research and Program of Russian Ministry of Education and Science.

References

1. E. Asher, "The interaction between magnetization and polarization: phenomenological symmetry consideration.," J. Phys. Soc. Jpn., **28**, 7 (1969).

2. T.H. O'Dell, *The Electrodynamics of Magnetoelectric Media*, North-Holland Publ. Company, Amsterdam, 1970, 304p.

3. M.I. Bichurin, V.M. Petrov, "Electric field influence on antiferromagnetic resonance spectrum in iron borate," Phys. Solid State, **29**, 2509 (1987).

4. J. Van den Boomgard, A.M.J.G. Van Run, J. Van Suchtelen, "Magnetoelectricity in piezoelectric-magnetostrictive composites," Ferroelectrics, **10**, 295 (1976).

5. R.E. Newnham, D.P. Skinner, L.E. Cross, "Connectivity and piezoelectric-pyroelectric composites," Mater. Res. Bull., **13**, 525 (1978).

6. G. Harshe, J.O. Dougherty, R.E. Newnham, "Theoretical modelling of multilayer magnetoelectric composites," Int. J. Appl. Electromagn. Mater., **4**, 145 (1993).

7. G. Harshe, J.P. Dougherty, R.E. Newnham, "Theoretical modelling of 3-0, 0-3 magnetoelectric composites," Int. J. Appl. Electromagn. Mater., **4**, 161 (1993).

8.   M.I. Bichurin, V.M. Petrov, G. Srinivasan, "Modelling of magnetoelectric effect in ferromagnetic/piezoelectric multilayer composites," Ferroelectrics, **280**, 165 (2002).

9.   J. Zhai, J.F. Li, D. Viehland, M.I. Bichurin, "The fundamental property of piezoelectric and magnetostrictive laminated composites," J. Appl. Phys. **101**, 014102 (2007).

10. M.I. Bichurin., V.M. Petrov, G. Srinivasan, "Theory of magnetoelectric effects in ferromagnetic ferroelectric layer composites," J. Appl. Phys, **92**, 7681 (2002).

11. M.I. Bichurin, V.M. Petrov, G. Srinivasan, "Theory of low-frequency magnetoelectric coupling in magnetostrictive-piezoelectric bilayers," Phys. Rev., **B68**, 054402 (2003).

12. M.I. Bichurin, G. Srinivasan, R. Hayes, V.M. Petrov, "Magnetoelectric effects in ferromagnetic/piezoelectric multilayer composites," Kluwer Series on NATO Advanced Research Workshop (Proceedings of the 5th International Conference on Magnetoelectric Interaction Phenomena in Crystals, MEIPIC-5, Sudak, Ukraine) 35 (2004).

13. V.M. Petrov, G. Srinivasan, M.I. Bichurin, T.A. Galkina, "Theory of magnetoelectric effect for bending modes in magnetostrictive-piezoelectric bilayers." J. Appl. Phys. **105**, 063911 (2009).

14. V.M. Petrov, M.I. Bichurin, A.S. Tatarenko, G. Srinivasan, "Effective parameters of bilayer ferrite-piezoelectric composite," Bulletin of NovSU: ser. "Technical sciences," **23**, 20 (2003) (in Russian).

15. M.I. Bichurin, V.M. Petrov, "Method of investigation for magnetoelectric composites," Abstracts of conf. on magnetoelectronics, IRE RAS, M. 1995, 125.

16. Ramirez A.P., "Interpretation of S-state ion EPR spectra," J. Phys. Condens. Matter, **9**, 8171–8174 (1997).

17. G. Srinivasan, E.T. Rasmussen, J. Gallegos, R. Srinivasan, Yu.I. Bokhan, V.M. Laletin, "Novel magnetoelectric bilayer and multilayer structures of magnetostrictive and piezoelectric oxides," Phys. Rev., **B64**, 214 (2001).

18. M.I. Bichurin, V.M. Petrov, *et al.* "Modeling of magnetoelectric effects in ferromagnetic/piezoelectric bulk composites," Kluwer Series on NATO Advanced Research Workshop (Proceedings of the 5th International Conference on Magnetoelectric Interaction Phenomena in Crystals, MEIPIC-5, Sudak, Ukraine) 65 (2004).

19. M.I. Bichurin, V.M. Petrov, I.A. Kornev, "Investigation of magnetoelectric interaction in composite," Ferroelectrics, **204**, 289 (1997).

20. V.M. Petrov, G. Srinivasan, U. Laletsin, M.I. Bichurin, D.S. Tuskov, N. Paddubnaya, "Magnetoelectric effects in porous ferromagnetic-piezoelectric bulk composites: experiment and theory," Phys. Rev., **B75**, 174422 (2007).

21. V.M. Petrov, M.I. Bichurin, V.M. Laletin, N.N. Paddubnaya, G. Srinivasan, "Modeling of magnetoelectric effects in ferromagnetic/piezoelectric bulk composites," Condens. Mater., abstract cond-mat/0401645, 2004. http://arxiv.org/abs/cond-mat/0401645

22. A. Novick, B. Berry, *Anelastic Relaxation in Crystalline Solids*, Academic Press, New York, 1972, 677p.

23. V. Daniel, *Dielectric Relaxation*, Academic Press, New York, 1967, 534p.

24. G.S. Radchenko, A.V. Turik, "Giant piezoelectric effect in layered ferroelectric-polymer composites," Phys. Solid State, **45**, 1759 (2003).

25. A.V. Turik, G.S. Radchenko, "Maxwell-wagner relaxation of elastic constants of layered polar dielectrics," Phys. Solid State, **45**, 1060 (2003).

26. V.M. Petrov, M.I. Bichurin, G. Srinivasan, "Maxwell-Wagner relaxation in magnetoelectric composites," Tech. Phys. Lett., **30**, 81 (2004).

27. V.M. Petrov, M.I. Bichurin, G. Srinivasan, J. Zhai, D. Viehland, "Dispersion characteristics for low-frequency magnetoelectric coefficients in bulk ferrite-piezoelectric composites," Solid State Commun., **142**, 515 (2007).

28. M.I. Bichurin, V.M. Petrov, I.A. Kornev, "A method for study of composite magnetoelectric properties," Abstracts of Int. Symp. "Ferro-piezoelectric mater. and their appl.," Moscow, Russia, 31 (1994) (in Russian).

29. V.M. Laletin, N. Paddubnaya, V.M. Petrov, M.I. Bichurin, "Theory of magnetoelectric effects in ferromagnetic/piezoelectric bulk composites," Bull. Am. Phys. Soc., 224 (2004).

30. O.V. Rybkov, V.M. Petrov, S.V. Averkin, G. Srinivasan, "Magnetoacoustic resonance in ferrite-piezoelectric bilayer structures subject

to exchange interaction," Bulletin of NovSU, **39**, 110 (2006) (in Russian).

31. V.M. Petrov, V.A. Nesterov, M.I. Bichurin, D.R. Buskunov, G. Srinivasan, "Relaxation magnetoelectric effect in bulk ferrite-piezoelectric composites," Proc. III Int. scientific and technical conf. "Problems of design and production of radioelectronic devices." Novopolotsk, Belorussiya, 26–28 May, 2004, 192 (in Russian).

32. V.M. Petrov, "Magnetoelectric susceptibility of multilayer ferrite-piezoelectric composite," Bulletin of NovSU: ser. "Natural and technical sciences," **26**, 19 (2004) (in Russian).

Chapter 3

Magnetoelectric Effect in the Electromechanical Resonance Range

M.I. Bichurin and V.M. Petrov

Institute of Electronic and Information Systems, Novgorod State University,
173003 *Veliky Novgorod, Russia*
Mirza.Bichurin@novsu.ru

In this chapter, we have presented a theory for the resonance enhancement of magnetoelectric (ME) interactions at frequencies corresponding to electromechanical resonance (EMR). Frequency dependence of ME voltage coefficients are obtained using the simultaneous solution of electrostatic, magnetostatic, and elastodynamic equations. The ME voltage coefficients are estimated from known material parameters (piezoelectric coupling, magnetostriction, elastic constants, etc.) of composite components. It is shown that the resonance enhancement of ME interactions is observed at frequencies corresponding to EMR and ME coupling in the EMR region exceeds the low-frequency value by more than an order of magnitude. It was found that the peak transverse ME coefficient at EMR is larger than the longitudinal one. The results of calculations obtained for a nickel-ferrite spinel–PZT composite are in good agreement with the experimental data.

Magnetoelectricity in Composites
Edited by Mirza I. Bichurin and Dwight Viehland
Copyright © 2012 Pan Stanford Publishing Pte. Ltd.
www.panstanford.com

The magnetoelectric (ME) effect in composites is caused by mechanically coupled magnetostrictive and piezoelectric subsystems: it is present in neither subsystem separately. Under magnetic field owing to the magnetostriction of the ferrite component, there are stresses which are elastically transmitted in the piezoelectric phase resulting in polarization changes via piezoelectricity. Because the ME effect in composites is due to mechanically coupled piezoelectric and magnetostrictive subsystems, it sharply increases in the vicinity of the electromechanical resonance (EMR) frequency [1, 2, 3, 6].

Mechanical oscillations of a ME composite can be induced either by alternating magnetic or electric fields. If the length of the electromagnetic wave exceeds the spatial size of the composite by some orders of magnitude, then it is possible to neglect gradients of the electric and magnetic fields within the sample volume. Therefore, based on elastodynamics and electrostatics, the equations of medium motion are governed by

$$\bar{\rho}\frac{\partial^2 u_i}{\partial t^2} = V\frac{\partial\,{}^pT_{ij}}{\partial x_j} + (1-V)\frac{\partial\,{}^mT_{ij}}{\partial x_j}, \tag{3.1}$$

where u_i is the displacement vector component, $\rho = V{}^p\rho + (1-V){}^m\rho$ is the average mass density, V is the ferroelectric volume fraction, ${}^p\rho$ and ${}^m\rho$, ${}^pT_{ij}$, and ${}^mT_{ij}$ are the densities and stress tensor components of ferroelectric and ferromagnet, correspondingly.

Simultaneous solution of elasticity equations and Eq. 3.1 by use of appropriate boundary conditions allows one to find the ME voltage coefficient. Since the solution depends on the composite shape and orientation of applied electric and magnetic fields, we consider some of the most general cases in this chapter.

3.1 NARROW COMPOSITE PLATE

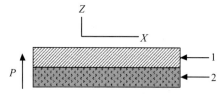

Figure 3.1 Scheme of bilayer of piezoelectric (1) and piezomagnetic (2) phases. The indicator specifies a direction of polarization.

First, let us consider a composite that has the form of a narrow plate which has a length L, as shown in Fig. 3.1 [5].

Tensorial expressions for the strain in the ferromagnetic layers ${}^mS_i$, and the strain ${}^pS_i$ and electrical displacement D in the ferroelectric layers have the form for the biasfield directed perpendicular to the sample plane (along z-axis):

$$
\begin{aligned}
{}^pS_1 &= {}^ps_{11}{}^pT_1 + {}^ps_{12}{}^pT_2 + {}^pd_{31}E_3, \\
{}^pS_2 &= {}^ps_{12}{}^pT_1 + {}^ps_{11}{}^pT_2 + {}^pd_{31}E_3,
\end{aligned}
\tag{3.2}
$$

$$
\begin{aligned}
{}^mS_1 &= {}^ms_{11}{}^mT_1 + {}^ms_{12}{}^mT_2 + {}^mq_{31}H_3, \\
{}^mS_2 &= {}^ms_{12}{}^mT_1 + {}^ms_{11}{}^mT_2 + {}^mq_{31}H_3,
\end{aligned}
\tag{3.3}
$$

$$
{}^pD_3 = {}^p\varepsilon_{33}E_3 + {}^pd_{31}\left({}^pT_1 + {}^pT_2\right),
\tag{3.4}
$$

where ${}^pT_i$ is the stress and ${}^ps_{ii}$ is the compliance of the ferroelectric at constant electric field, ${}^mT_i$ is the stress and ${}^ms_{ii}$ is the compliance of the ferromagnetic at constant magnetic field, ${}^p\varepsilon_{33}$ is the relevant component of the electrical permittivity, ${}^pd_{31}$ is the piezoelectric coefficient of the ferroelectric, ${}^mq_{ij}$ is the piezomagnetic coefficient of ferromagnetic, E_3 and H_3 are ac electric and magnetic fields. Close to EMR we can assume ${}^mT_1 \gg {}^mT_2$ and ${}^pT_1 \gg {}^pT_2$ (axis 1 is directed along the plate length) such that ${}^mT_2$ and ${}^pT_2$ may be ignored.

Expressions for the stress components ${}^pT_1$ and ${}^mT_1$ can be found from Eqs. 3.2 and 3.3. Substituting these expressions into Eq. 3.1 yields a differential equation for u_x. Assuming harmonic motion along x, we get the solution of this equation in the form:

$$
u_x = A\cos(kx) + B\sin(kx),
\tag{3.5}
$$

where $k = \omega\sqrt{-\rho\left[\dfrac{V}{{}^ps_{11}} + \dfrac{1-V}{{}^ms_{11}}\right]^{-1}}$ and ω is angular frequency. The integration constants A and B can be found from the boundary conditions. Assuming that the sample surfaces at $x = 0$ and $x = L$ are free from external stresses, we have the following boundary conditions:

$$
V{}^pT_1 + (1-V){}^mT_1 = 0, \text{ at } x = 0 \text{ and } x = L.
\tag{3.6}
$$

For determining the ME voltage coefficient, we use the open circuit condition:

$$\int_0^L {}^P D_3 dx = 0.$$

(3.7)

Substituting Eq. 3.4 into Eq. 3.7 and taking into account Eqs. 3.5 and 3.6, we can derive

$$\alpha_{E,33} = \frac{2\,{}^P d_{31}\,{}^m q_{31}\,{}^P s_{11} V(1-V)\tan(^{kL}\!\!/_2)}{s_1({}^P d_{31}^2 - {}^P s_{11}\,{}^P \varepsilon_{33})kL - 2\,{}^P d_{31}^2 V\,{}^m s_{11}\tan(^{kL}\!\!/_2)},$$

(3.8)

where $s_1 = V^m s_{11} + (1-V)^P s_{11}$.

For transverse field orientation, magnetic induction B has the only component B_1 which should obey the condition $\partial B_1/\partial x = 0$ since B is divergence free. For this case, Eq. 3.3 should be written in the more convenient form:

$${}^m S_1 = {}^m s^B_{11}\,{}^m T_1 + {}^m g_{11} B_1,$$

(3.9)

where ${}^m s^B_{11}$ is compliance at constant magnetic induction and ${}^m g_{11}$ is piezomagnetic coefficient, ${}^m g_{11} = \partial\,{}^m S_1/\partial B_1$.

In a similar manner as the calculation above, the transverse ME voltage coefficient can shown to be

$$\alpha_{E,31} = \frac{2\,{}^P d_{31}\,{}^m g_{11}\mu_{eff}\,{}^P s_{11} V(1-V)\tan(^{kL}\!\!/_2)}{s_2({}^P d_{31}^2 - {}^P s_{11}\,{}^P \varepsilon_{33})kL - 2\,{}^P d_{31}^2 V\,{}^m s^B_{11}\tan(^{kL}\!\!/_2)},$$

(3.10)

where $s_2 = V^m s^B_{11} + (1-V)^P s_{11}$. Effective permeability μ_{eff} can be found from constitutive equation

$$H_1 = -{}^m g_{11}\,{}^m T_1 + B_1/{}^m \mu_{11},$$

(3.11)

where ${}^m \mu_{11}$ is permeability of magnetic phase.

Expressing ${}^m T_1$ from Eq. 3.3 and substituting it into Eq. 3.11 enables finding μ_{eff} with the use of Eq. 3.5 after integrating over the sample length.

$$\mu_{eff} = \frac{s_2 \, {}^m s_{11}^B \, {}^m \mu_{11} kL}{({}^m s_{11}^B + {}^m g_{11}^2 \, {}^m \mu_{11})kLs_2 + 2 \, {}^m g_{11}^2 \, {}^m \mu_{11} \, {}^p s_{11}(1-V)\tan(kL/2)}.$$

(3.12)

As one can see from Eqs. 3.8 and 3.10 that the value of the ME coefficient under applied fields is directly proportional to the product of piezoelectric d_{31} and piezomagnetic q_{31} or g_{11} modules. Bear in mind, that in reality, there are always loss factors that must be included, even in "perfect" materials if for no other reason than losses associated with electrical contacts. Said loss factors define the width of the resonant line, limiting the peak value of the ME coefficient. The width of the resonant peak can be varied through attenuation coefficients. Such coefficients are present in k and ω [1], as they are both complex parameters. We shall use a complex frequency $\omega(1 + i/Q)$ with Q to be determined experimentally.

The roots of the denominator in Eqs. 3.8 and 3.10 define the maxima in the frequency dependence of the ME voltage coefficient. In Figs. 3.2 and 3.3, the frequency dependence of the transverse and longitudinal ME voltage coefficients is shown for the bilayer of nickel ferrite (NFO) and lead zirconate titanate (PZT). In these figures, the resonance peaks caused by oscillations along the x-axes can be seen. The maximum value of the ME coefficient (5400 mV (cmOe)) is observed for the transverse field orientation, whereas the value at frequency of 100 Hz is 144 mV/(cmOe). Thus, the resonant value of the ME coefficient exceeds the low-frequency value by a factor of about 40. For the longitudinal field orientation, the magnitude of the ME effect is smaller by one order of magnitude. This is explained by the fact that, for the longitudinal field orientation, the ME effect is significantly affected by demagnetizing fields.

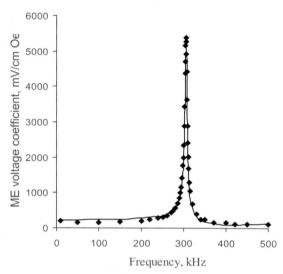

Figure 3.2 Frequency dependence of transverse ME voltage coefficient for the bilayer of NFO and PZT of 7.3 mm length. $Q = 250$ and PZT volume fraction is 0.6. There is a good agreement between calculation (solid line) and data (points).

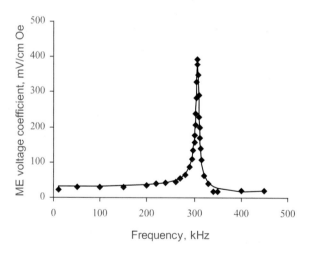

Figure 3.3 Frequency dependence of longitudinal ME voltage coefficient for the bilayer of NFO and PZT of 7.3 mm length. $Q = 105$ and PZT volume fraction is 0.6. There is a good agreement between calculation (solid line) and data (points).

3.2 DISC-SHAPED BILAYER

Let us consider now a ferrite–piezoelectric disk-shaped composite of radius R and thickness d, which has thin metal electrodes deposited on bottom and top surfaces, as shown in Fig. 3.4.

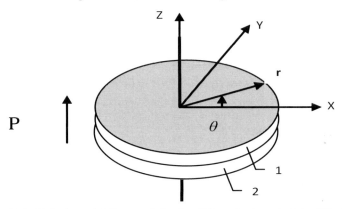

Fig. 3.4. Disk-shaped bilayer of ferrite (1) and piezoelectric (2). The indicator notes a direction of polarization.

Assume that the sample is poled normal to the plane of the electrodes, along the z-axis. DC and ac magnetic fields can be directed either along the normal or in the plane of the contacts, which distinguishes the cases of longitudinal and transverse field orientations, respectively. Due to magnetostriction, application of an ac magnetic field will excite both thickness and radial oscillations. In what follows we consider the low-frequency radial oscillations. Radial oscillations will have a notably lower frequency than thickness mode ones, simple because of the geometrical size limitations of the thickness mode.

The disc is supposed to be thin (i.e., $d \ll R$). Since the surfaces are free from external forces, the normal components of the stress tensors will be zero. For a thin disc, it is reasonable to assume that the component of the stress tensor T_3 is zero not only on the surfaces, but also in the volume of the disc. In addition, since the top and bottom bases of the disc are equipotential surfaces, only the z component of the electric field vector is non-zero. Then, the constitutive equations are determined by Eqs. 3.2–3.4 for longitudinal field orientation.

It is convenient to take advantage of the symmetry of a disk by use of a cylindrical coordinate system (z, r, and θ). In this case, the strain and stress tensor components need to be transformed in by the following relations [7]:

$$^iS_1 = {}^iS_{rr}\cos^2(\theta) - 2^iS_{r\theta}\sin(\theta)\cos(\theta) + {}^iS_{\theta\theta}\sin^2(\theta), \qquad (3.13)$$

$$^iS_2 = {}^iS_{rr}\sin^2(\theta) + 2^iS_{r\theta}\sin(\theta)\cos(\theta) + {}^iS_{\theta\theta}\cos^2(\theta), \qquad (3.14)$$

$$^iT_{rr} = {}^iT_1\cos^2(\theta) + 2^iT_6\sin(\theta)\cos(\theta) + {}^iT_2\sin^2(\theta), \qquad (3.15)$$

$$^iT_{\theta\theta} = {}^iT_1\sin^2(\theta) - 2^iT_6\sin(\theta)\cos(\theta) + {}^iT_2\cos^2(\theta), \qquad (3.16)$$

$$^iT_{r\theta} = \sin(\theta)\cos(\theta)[{}^iT_2 - {}^iT_1] + [\cos^2(\theta) - \sin^2(\theta)]^iT_6, \qquad (3.17)$$

where $i = p$ or m. The components of the strain tensor in cylindrical coordinates are defined via the vector displacement u, as follows:

$$^iS_{rr} = \partial u_r / \partial r, \qquad (3.18)$$

$$^iS_{\theta\theta} = u_r / r, \qquad (3.19)$$

$$^iS_{r\theta} = (1/r)\, \partial u_r / \partial \theta. \qquad (3.20)$$

The other necessary equation is the equation of medium motion transformed into cylindrical coordinates:

$$-\bar{\rho}\omega^2 u_r = V\left(\frac{\partial\, ^pT_{rr}}{\partial r} + \frac{\partial\, ^pT_{r\theta}}{r\partial\theta} + \frac{^pT_{rr} - {}^pT_{\theta\theta}}{r}\right)$$
$$+ (1-V)\left(\frac{\partial\, ^mT_{rr}}{\partial r} + \frac{\partial\, ^mT_{r\theta}}{r\partial\theta} + \frac{^mT_{rr} - {}^mT_{\theta\theta}}{r}\right) \qquad (3.21)$$

Solving the elasticity equations for stress components and substituting the found expressions into Eq. 3.21, we can obtain the equation for the radial displacements. However, the form of this equation depends on the electric and magnetic fields orientations. Next, we consider the ME coupling for longitudinal and transverse fields orientations separately.

3.2.1 Longitudinal Orientation of Electric and Magnetic Fields

In this case, the longitudinal orientation of dc and ac magnetic fields coincides with the direction of polarization. Taking into account the axial symmetry of disk-shaped sample, one can get the following expressions for nonvanishing strain components [4]:

$$
\begin{aligned}
{}^{p}S_{rr} &= {}^{p}s_{11}\,{}^{p}T_{rr} + {}^{p}s_{12}\,{}^{p}T_{\theta\theta} + {}^{p}d_{31}E_{3}, \\
{}^{m}S_{rr} &= {}^{m}s_{11}\,{}^{m}T_{rr} + {}^{m}s_{12}\,{}^{m}T_{\theta\theta} + {}^{m}q_{31}H_{3}, \\
{}^{p}S_{\theta\theta} &= {}^{p}s_{12}\,{}^{p}T_{rr} + {}^{p}s_{11}\,{}^{p}T_{\theta\theta} + {}^{p}d_{31}E_{3}, \\
{}^{m}S_{\theta\theta} &= {}^{m}s_{12}\,{}^{m}T_{rr} + {}^{m}s_{11}\,{}^{m}T_{\theta\theta} + {}^{m}q_{31}H_{3}
\end{aligned}
\tag{3.22}
$$

Solving Eqs. 3.22 for stress components and substituting these expressions into Eq. 3.21, we get the equation of motion in the form:

$$
-\omega^{2}\bar{\rho}u_{r} = \left[\frac{V}{{}^{p}s_{11}(1 - {}^{p}v^{2})} + \frac{1 - V}{{}^{m}s_{11}(1 - {}^{m}v^{2})} \right]\left(\frac{\partial^{2}u_{r}}{\partial r^{2}} + \frac{\partial u_{r}}{r\partial r} - \frac{u_{r}}{r^{2}} \right)
\tag{3.23}
$$

where ${}^{p}v$ and ${}^{m}v$ are Poisson's ratios for piezoelectric and magnetostrictive phases.

The solution of Eq. 3.23 is a linear combination of Bessel function of the first and second sort:

$$
u_{r} = c_{1}J_{1}(kr) + c_{2}Y_{1}(kr),
\tag{3.24}
$$

where $k = \omega\sqrt{\bar{\rho}\left[\dfrac{V}{{}^{p}s_{11}(1 - {}^{p}v^{2})} + \dfrac{1 - V}{{}^{m}s_{11}(1 - {}^{m}v^{2})} \right]^{-1}}$.

The constants of integration c_{1} and c_{2} should be found from boundary conditions: $u_{r} = 0$ at $r = 0$ and $T_{rr} = 0$ at $r = R$.

For obtaining the ME voltage coefficient, we use the open-circuit condition

$$
\int_{0}^{R} rdr \int_{0}^{2\pi} d\theta D_{3} = 0,
\tag{3.25}
$$

where electric induction is determined by the constitutive equation

$$D_3 = {}^Pd_{31}({}^PT_{rr} + {}^PT_{\theta\theta}) + {}^P\varepsilon_{33}E_3 .\qquad(3.26)$$

Substituting Eq. 3.24 into Eqs. 3.18 and 3.19 and then into 3.22, we find the expressions for stress components. Once the stress components are determined, the ME voltage coefficient can be found from Eq. 3.25.

$$\alpha_{E,L} = -\frac{2(1+v)(1-V){}^Ps_{11}J_1(kR){}^Pd_{31}{}^mq_{31}}{(1-v){}^Ps_{11}[aJ_0(kR)-(1-v)s_1J_1(kR)]{}^P\varepsilon_{33}+2[aJ_0(kR)-s_3J_1(kR)]{}^Pd_{31}^2}$$

$$(3.27)$$

where $a = kRs_1$, $s_1 = V^ms_{11} + (1-V){}^Ps_{11}$, $s_3 = (1-v)(1-V){}^Ps_{11} + 2V^ms_{11}$. It should be noted that ${}^mv$ is assumed to equal ${}^Pv = v$ for simplicity of Eq. 3.27.

3.2.2 Transverse Orientation of Electric and Magnetic Fields

In this case, the in-plane dc and ac magnetic field vectors are perpendicular to the electric field vector field which is along the z-axis. Eqs. 3.22 should be replaced by

$$
\begin{aligned}
{}^PS_{rr} &= {}^Ps_{11}{}^PT_{rr} + {}^Ps_{12}{}^PT_{\theta\theta} + {}^Pd_{31}E_3, \\
{}^mS_{rr} &= {}^ms_{11}^B{}^mT_{rr} + {}^ms_{12}^B{}^mT_{\theta\theta} + [{}^mg_{11}\cos^2(\theta)+{}^mg_{12}\sin^2(\theta)]B_1, \\
{}^PS_{\theta\theta} &= {}^Ps_{12}{}^PT_{rr} + {}^Ps_{11}{}^PT_{\theta\theta} + {}^Pd_{31}E_3, \\
{}^mS_{\theta\theta} &= {}^ms_{12}^B{}^mT_{rr} + {}^ms_{11}^B{}^mT_{\theta\theta} + [{}^mg_{12}\cos^2(\theta)+{}^mg_{11}\sin^2(\theta)]B_1, \\
{}^PS_{r\theta} &= ({}^Ps_{11} - {}^Ps_{12}){}^PT_{r\theta}, \\
{}^mS_{r\theta} &= ({}^ms_{11}^B - {}^ms_{12}^B){}^mT_{r\theta} - \frac{1}{2}\sin(2\theta)({}^mg_{11}-{}^mg_{12})B_1
\end{aligned}
$$

$$(3.28)$$

Solving Eqs. 3.28 for stress components and substituting these expressions into Eq. 3.13 results in the following form of equation of media motion:

$$-\omega^2 \bar{\rho} u_r = \left[\frac{V}{{}^p s_{11}(1-{}^p v^2)} + \frac{1-V}{{}^m s_{11}(1-{}^m v^2)} \right] \left(\frac{\partial^2 u_r}{\partial r^2} + \frac{\partial u_r}{r \partial r} - \frac{u_r}{r^2} \right)$$

$$+ \left[\frac{V}{{}^p s_{11}(1+{}^p v)} + \frac{1-V}{{}^m s_{11}(1+{}^m v)} \right] \frac{\partial^2 u_r}{r^2 \partial \theta^2} \qquad (3.29)$$

The solution of Eq. 3.29 corresponding to the radial mode is defined

by Eq. 3.24 with $k = \omega \sqrt{\bar{\rho} \left[\dfrac{V}{{}^p s_{11}(1-{}^p v^2)} + \dfrac{1-V}{{}^m s_{11}^B (1-{}^m v^2)} \right]^{-1}}$. The

constants of integration c_1 and c_2 should be found from boundary

conditions: $u_r = 0$ at $r = 0$ and $\displaystyle\int_0^{2\pi} T_{rr} d\theta = 0$ at $r = R$.

Substituting Eq. 3.24 into Eqs. 3.18 and 3.19 and then into 3.28, we find the expressions for stress components. Once the stress components are determined, the ME voltage coefficient can be found from Eq. 3.25.

$$\alpha_{E,T} = -\frac{(1+v)(1-V)\,{}^p s_{11} J_1(kR)\,{}^p d_{31}({}^m g_{11} + {}^m g_{12})\mu_{eff}}{(1-v)\,{}^p s_{11}[aJ_0(kR) - (1-v)s_1 J_1(kR)]\,{}^p \varepsilon_{33} + 2[aJ_0(kR) - s_3 J_1(kR)]\,{}^p d_{31}^2}$$

$$(3.30)$$

where $a = kRs_1$, $s_1 = V^m s^B{}_{11} + (1-V)^p s_{11}$, $s_3 = (1-v)(1-V)^p s_{11} + 2V^m s^B{}_{11}$. Effective permeability μ_{eff} can be found similarly to section 3.1. It should be noted that ${}^m v$ is assumed to equal ${}^p v$ for simplicity of Eq. 3.30.

Experimental investigations of the ME effect have been performed for bilayer of NFO and PZT. The samples were discs with the radius of $R = 4.7$ mm, and were poled by dc electric field $E = 4$ kV/mm for 3 hours at 80°C. The bias field dependence of low-frequency ME effect was measured in the beginning. Measurements were done for both longitudinal (see Fig. 3.5) and transverse (see Fig. 3.6) orientations of electric and magnetic fields in the EMR region. Experimental data and corresponding theoretical estimates [5] based on Eqs. 3.27 and 3.30 are given in Figs. 3.5 and 3.6, for longitudinal and transverse fields respectively. Calculations were performed using the following values of the materials parameters: for NFO $^m s_{11} = 6.5 \cdot 10^{-12}$ m²/м, $^m s_{12} = -2.4 \cdot 10^{-12}$ m²/м, $^m q_{31} = 70 \cdot 10^{-12}$ m/A, $^m q_{11} = -430 \cdot 10^{-12}$ m/A, $^m q_{12} = 125 \cdot 10^{-12}$ m/A, $^m \varepsilon_{33}/\varepsilon_0 = 10$; and for PZT $^p s_{11} = 15.3 \cdot 10^{-12}$ m²/м, $^p s_{12} = 5 \cdot 10^{-12}$ m²/м, $^p d_{31} = -175 \cdot 10^{-12}$ m/B, $^p \varepsilon_{33}/\varepsilon_0 = 1750$. The attenuation parameter was empirically defined by the line-width of the EMR peak at its half-maximum point.

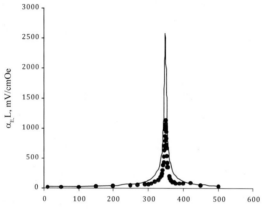

Figure 3.5 Frequency dependence of magnetoelectric voltage coefficient at longitudinal orientation of fields. Solid line — theory, points — experiment.

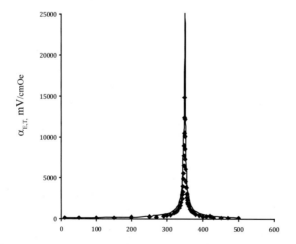

Figure 3.6 Frequency dependence of magnetoelectric voltage coefficient at transverse orientation of fields. Solid line — theory, points — experiment.

As can be seen in these figures, there is a relatively good agreement between theory and experimental data. The maximum value of the transverse ME coefficient for the disc-shaped composites was about 15 V/cmOe at resonance, whereas the low-frequency value was only about 0.16 V/cmOe. The coefficient of attenuation was found to be smaller for transverse fields compared to longitudinal ones. This can be accounted for by higher energy losses for longitudinal fields orientation due to Eddy currents in the electrodes. In general, it was found that the transverse ME coefficient was larger than the longitudinal one. As mentioned earlier, this is also due to the fact that the demagnetization field appears at longitudinal orientations that reduces the piezomagnetic modulus.

3.3 CONCLUSIONS

In this chapter, we have presented a theory for the resonance enhancement of ME interactions at frequencies corresponding to EMR. Frequency dependence for longitudinal and transverse ME voltage coefficients are obtained using the simultaneous solution of electrostatic, magnetostatic, and elastodynamic equations. The ME voltage coefficients are estimated from known material parameters (piezoelectric coupling, magnetostriction, elastic constants, etc.) of composite components.

It is shown that the ME coupling in the EMR region exceeds the low-frequency value by more than an order of magnitude. It was found that the peak transverse ME coefficient at EMR is larger than the longitudinal one. This is accounted for (i) by higher energy losses for longitudinal fields orientation due to eddy currents in the electrodes and (ii) by influence of demagnetization field that appears at longitudinal orientations that reduces the piezomagnetic modulus.

The results of calculations obtained for a NFO spinel–PZT composite are in good agreement with the experimental data.

Acknowledgments

This work was supported by the Russian Foundation for Basic Research and Program of Russian Ministry of Education and Science.

References

1. R.P. Santoro, R.E. Newnham, "Survey of magnetoelectric materials," Technical Report AFML TR-66–327, Air Force Materials Lab., Ohio, 1966.

2. M.I. Bichurin, D.A. Filippov, V.M. Petrov, G. Srinivasan, "Magneto-piezoelectric and electro-piezomagnetic effects in composites," Physics of electric materials: Proc. Int. Conf. 1–4 October 2002, Kaluga, Russia, 309 (in Russian).

3. M.I. Bichurin, V.M. Petrov, D.A. Filippov, G. Srinivasan, C.-W. Nan, "Magnetoelectric effect in ferrite-piezoelectric composites," Proc. 11 Int. Conf. on Electromagnetics, Electrotechnology and Electromaterial Science, Crimea, Alushta, 147 (2006) (in Russian).

4. M.I. Bichurin, D.A. Filippov, V.M. Petrov, V.M. Laletsin, N.N. Paddubnaya, G. Srinivasan, "Resonance magnetoelectric effects in layered magnetostrictive-piezoelectric composites," Phys. Rev., **B68**, 132408 (2003).

5. M.I. Bichurin, V.M. Petrov, S.V. Averkin, A.V. Filippov, "Electromechanical Resonance in Magnetoelectric Layered Structures," Phys. Solid State, **52**, 2116 (2010).

6. M.I. Bichurin, V.M. Petrov, D.A. Filippov, G. Srinivasan, "Modeling of magneto-piezoelectric and electro-piezomagnetic effects in composites," Proc. IV All-Russian scientific internet-conf.: "Computer and mathematical modeling in science and technology" (CMM-4)/ TSU, Tambov, **15**, 68 (2002) (in Russian).

7. W.P. Mazon, "Electrostrictive effect in barium titanate ceramics," Phys. Rev., **74**, 1134 (1948).

Chapter 4

Magnetoelectric Effect and Green's Function Method

Ce-Wen Nan

Department of Materials Science and Engineering, Tsinghua University,
Beijing 100084, China
cwnan@tsinghua.edu.cn

The primary task of this chapter is to express the effective coupling properties of magnetoelectric (ME) composite in a convenient matrix formulation by using the successful Green's function technique. This approach described the influence of the composite microstructure (e.g., phase volume fraction, phase connectivity, aspect ratio of the phase particles) on the effective properties of ME composites. In addition, the giant ME effect in composites containing Terfenol-D is predicted. In the ME nanostructures such as nanocomposite thin films, the ME coupling is still strain-mediated in most cases. The strain-mediated ME coupling in the nanostructures can also be modeled by the Green's function technique. The physics-based approach for the ME composites remains to be further generalized to treat the dynamic ME behavior of these ME composites including bulk and nanostructures.

As mentioned above, magnetoelectric (ME) composites [1] made by combining piezoelectric and magnetic substances together have drawn significant interest in recent years due to their multifunctionality, in which the coupling interaction between

Magnetoelectricity in Composites
Edited by Mirza I. Bichurin and Dwight Viehland
Copyright © 2012 Pan Stanford Publishing Pte. Ltd.
www.panstanford.com

piezoelectric and magnetic substances could produce a large ME response [2, 3] (e.g., several orders of magnitude higher than that in those single-phase ME materials so far available) at room temperature. These ME composites provide opportunities for potential applications as multifunctional devices such as magnetic–electric transducers, actuators, and sensors.

The ME effect in composite materials is known as a product tensor property [1], which results from the cross interaction between different ordering of the two phases in the composite. Neither the piezoelectric nor magnetic phase has the ME effect, but composites of these two phases have remarkable ME effect. Thus, the ME effect is a result of the product of the magnetostrictive effect (magnetic/mechanical effect) in the magnetic phase and the piezoelectric effect (mechanical/electrical effect) in the piezoelectric one, namely [4]

$$\text{ME}_\text{H} \text{ effect} = \frac{\text{magnetic}}{\text{mechanical}} \times \frac{\text{mechanical}}{\text{electric}} \qquad (4.1a)$$

$$\text{ME}_\text{E} \text{ effect} = \frac{\text{electric}}{\text{mechanical}} \times \frac{\text{mechanical}}{\text{magnetic}}. \qquad (4.1b)$$

This is a coupled electrical and magnetic phenomenon via elastic interaction. That is, for the ME_H effect, when a magnetic field is applied to a composite, the magnetic phase changes its shape magnetostrictively. The strain is then passed along to the piezoelectric phase, resulting in an electric polarization. Thus, the ME effect in composites is extrinsic, depending on the composite microstructure and coupling interaction across magnetic–piezoelectric interfaces.

Due to the technologically viable ME response observed in multiferroic ME composites above room temperature, various ME composites in different systems have been investigated in recent years, including (1) bulk ceramic ME composites of piezoelectric ceramics and ferrites [5, 6]; (2) two-phase ME composites of magnetic alloys and piezoelectric materials [7, 8]; (3) three-phase ME composites [9]; and (4) thin films [10] (nanostructured composites) of ferroelectric and magnetic oxides. Using the concept of phase connectivity, we can describe the structures of a two-phase composite using the notations 0–3, 2–2, and 1–3, etc., in which each number denotes the connectivity of the respective phase. For example, a 0–3 type particulate composite means one phase particles (denoted by 0) embedded in the matrix of another phase (denoted by 3). So far, 0–3, 3–3, 2–2, and 1–3 type structured ME composites of ferroelectric and magnetic phases have been developed. Some prototype ME devices based on the ME composites have been proposed due to their large ME effect at room temperature [2, 3]. In this chapter, we will focus on a physically based approach to these ME composites from bulk materials to thin films.

4.1 BULK CERAMIC COMPOSITES

4.1.1 Green's Function Technique

As for piezoelectric composites, the ME composites could have various connectivity schemes, but the common connectivity schemes are 0–3 type particulate composites of piezoelectric and magnetic oxide grains, 2–2 type laminate ceramic composites consisting of piezoelectric and magnetic oxide layers, and 1–3 type fiber composites with fibers of one phase embedded in the matrix of another phase, as shown in Fig. 4.1. $BaTiO_3$, PZT, $Pb(MgNb)O_3$–$PbTiO_3$, etc. are usually chosen as the piezoelectric ceramic phase, and ferrites usually as the magnetic phase (Table 4.1).

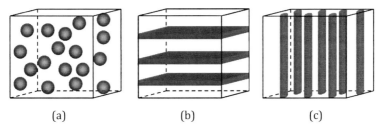

(a)   (b)   (c)

Figure 4.1 Schematic illustration of three bulk composites with the three common connectivity schemes: (a) 0–3 particulate composite, (b) 2–2 laminate composite, and (c) 1–3 fiber/rod composite.

The constitutive equations for describing coupling mechanical–electric–magnetic response in the ME composites, in a linear approximation, can be written by direct notation for tensors as

$$\sigma = c\,S - e^T E - q^T H$$
$$D = e\,S + \varepsilon\,E + \alpha\,H \qquad\qquad (4.2)$$
$$B = q\,S + \alpha^T E + \mu\,H,$$

where σ, S, D, E, B, and H are the stress, strain, electric displacement, electric field, magnetic induction, and magnetic field, respectively; c, ε, and μ are, respectively, the stiffness, dielectric constant, and permeability; e and q are the piezoelectric and piezomagnetic coefficients, respectively; and α is the ME coefficient. The superscript T means the transpose of the tensor. The tensors c, e, q, ε, μ, and α are (6×6), (3×6), (3×6), (3×3), (3×3), and (3×3) matrices, respectively, by means of the compressive representation. For the composite, all these tensors are local quantities depending on the spatial position $\mathbf{x}$. The effective properties of the composite can be defined in terms of the averaged fields, namely,

$$\langle\sigma\rangle = c^* \langle S\rangle - e^{T*} \langle E\rangle - q^{T*} \langle H\rangle$$
$$\langle D\rangle = e^* \langle S\rangle + \varepsilon^* \langle E\rangle + \alpha^* \langle H\rangle \qquad\qquad (4.3)$$
$$\langle B\rangle = q^* \langle S\rangle + \alpha^{T*} \langle E\rangle + \mu^* \langle H\rangle,$$

where angular brackets denote the microstructural average and quantities with asterisks represent those of the composite. Therefore, the problem of evaluating the effective response of the material essentially consists of the determination of the field variables within it under certain specified boundary conditions, followed by the performance of the averages [4].

Because of spatial variations in the constitutive behavior in the composite with position, the local constitutive coefficients X (e.g., c, ε, and μ) can be written as a variation about a mean value, e.g.,

$$X = X^o + X', \qquad\qquad (4.4)$$

where the first terms X^o denoted by superscript o represent the constitutive constants of a homogeneous reference medium and the second terms are the spatial fluctuations of the first.

Now let the composite be subjected on its external surface S to homogeneous mechanical–electric–magnetic boundary conditions, i.e.,

$$u_i(S) = S^o_{ij} x_j = u^o_i, \quad \phi(S) = -E^o_i x_i = \phi^o, \quad \varphi(S) = -H^o_i x_i = \varphi^o,$$

$$(4.5)$$

where u_i, ϕ, and φ, respectively, denote the elastic displacement, electric potentials, and magnetic potential. Consider a state of static equilibrium in the absence of body forces and free electric charges so that

$$\sigma_{ij,j} = 0, \, D_{i,i} = 0, \, B_{i,i} = 0, \qquad (4.6)$$

where the commas in the subscripts denote partial differentiation with respect to x_j. These are nonlinearly coupling equilibrium equations.

Further, by solving the equilibrium equations under the boundary conditions in terms of the Green's function technique, the local fields within the composite can be obtained as [4, 11]

$$\begin{pmatrix} S \\ E \\ H \end{pmatrix} = \begin{pmatrix} T^{11} & -T^{12} & -T^{13} \\ T^{21} & T^{22} & T^{23} \\ T^{31} & T^{32} & T^{33} \end{pmatrix} \begin{pmatrix} S^o \\ E^o \\ H^o \end{pmatrix} \qquad (4.7a)$$

$$T^{11} = \left[(T^{66})^{-1} + G^u q^T [I - G^v (\mu - \mu^o)]^{-1} G^v q \right]^{-1}$$
$$T^{12} = T^{11} G^u e^T [I - G^\phi (\varepsilon - \varepsilon^o)]^{-1} \qquad (4.7b)$$
$$T^{13} = T^{11} G^u q^T [I - G^v (\mu - \mu^o)]^{-1}$$

$$T^{21} = [I - G^\phi (\varepsilon - \varepsilon^o)]^{-1} G^\phi e T^{11}$$
$$T^{22} = [I - G^\phi (\varepsilon - \varepsilon^o)]^{-1} (I + G^\phi e T^{12}) \qquad (4.7c)$$
$$T^{23} = [I - G^\phi (\varepsilon - \varepsilon^o)]^{-1} G^\phi e T^{13}$$

$$T^{31} = [I - G^v (\mu - \mu^o)]^{-1} G^v q T^{11}$$
$$T^{32} = [I - G^v (\mu - \mu^o)]^{-1} G^v q T^{12}$$
$$T^{33} = [I - G^v (\mu - \mu^o)]^{-1} (I + G^v q T^{13})$$
$$T^{66} = \left[I - G^u (c - c^o)^{-1} + G^u e^T [I - G^\phi (\varepsilon - \varepsilon^o)]^{-1} G^\phi e \right]^{-1}.$$

These T^{ij} are so-called t-matrix tensors. For the piezoelectric phase (e.g., $BaTiO_3$ and PZT) in the composites, $q = 0$ and $\alpha = 0$; and for the magnetic phase (e.g., Co-ferrites and Ni-ferrites) in the composites, $e = 0$ and $\alpha = 0$. Thus, by averaging these solutions for local field quantities and eliminating S^o, E^o, and H^o from them, we obtain the general solutions to overall effective properties of the composite

$$C^* = \langle CT^{11} - e^T T^{21} - q^T T^{31} \rangle A^{11} + \langle CT^{12} + e^T T^{22} \rangle A^{12}$$
$$+ \langle CT^{13} + q^T T^{33} \rangle A^{13}$$

$$e^{T*} = \langle (C - C^*)T^{12} + e^T T^{22} \rangle \langle T^{22} \rangle^{-1}$$

$$q^{T*} = \langle (C - C^*)T^{13} + q^T T^{33} \rangle \langle T^{33} \rangle^{-1}$$

$$\varepsilon^* = \langle (e^* - e)T^{12} + \varepsilon T^{22} \rangle \langle T^{22} \rangle^{-1} \tag{4.8a}$$

$$\mu^* = \langle (q^* - q)T^{13} + \mu T^{33} \rangle \langle T^{33} \rangle^{-1}$$

$$\alpha^* = \langle (e^* - e)T^{13} \rangle \langle T^{33} \rangle^{-1} = \langle (q^* - q)T^{12} \rangle \langle T^{22} \rangle^{-1},$$

where

$$A^{11} = \left[\langle T^{11} \rangle + \langle T^{12} \rangle \langle T^{22} \rangle^{-1} \langle T^{21} \rangle + \langle T^{13} \rangle \langle T^{33} \rangle^{-1} \langle T^{31} \rangle \right]^{-1}$$

$$A^{12} = \langle T^{22} \rangle^{-1} \langle T^{21} \rangle A^{11} \tag{4.8b}$$

$$A^{13} = \langle T^{33} \rangle^{-1} \langle T^{31} \rangle A^{11}.$$

These results are quite general and independent of the models assumed for the topology of the phases in the composite. These complicated, coupled expressions lack closed form solutions, but their convenient matrix formulations are particularly suited for easily numerical calculations of the overall effective properties. As seen, for their composites, the effective ME coefficient $\alpha^* \neq 0$, which depends on details of the composite microstructures, i.e., component phase properties, volume fraction, grain shape, phase connectivity, etc.

4.1.2 Some Approximations

In some special cases, these complicated, coupled expressions can reduce to explicit closed form solutions. As an example, firstly consider a 1–3 type composite (Fig. 4.1) with piezoelectric (or magnetic) rods aligned in a magnetic (or piezoelectric) matrix. Let the piezoelectric phase be poled along x_3 axis of the composite. Thus, the composite has a ∞mm symmetry with ∞ denoting the x_3 axis. The magnetic field is also along the symmetric x_3 axis. In the limit case that the aspect ratio p of the rods approaches infinite, Eq. 4.8a gives the following simple expression for the ME coefficient along the symmetric x_3 axis

$$\alpha_{33}^* = -f\,\frac{e_{31}^* q_{31}}{{}^m k + m^o} = -(1-f)\frac{q_{31}^* e_{31}}{{}^p k + m^o}, \qquad (4.9)$$

where f is the volume fraction of the magnetic phase, $k = (c_{11} + c_{12})/2$, and $m = (c_{11} - c_{12})/2$ are, respectively, transverse in-plane bulk modulus and transverse shear modulus (the superscripts "m" and "p" denoting the magnetic and piezoelectric phases; the superscript o denoting a homogeneous reference medium); q_{31} and e_{31} are, respectively, piezomagnetic and piezoelectric coefficients. Different approximations pertain and can be easily obtained from the general solution 4.9, depending on the choice made for m^o of the homogeneous reference medium. An intuitive and common choice is to take the host matrix phase as the reference medium. In essence it is a non-self-consistent approximation (NSCA) which is generally valid for matrix-based composites such as a 0–3 type particulate microstructure. Naturally, for the choice $m^o = m^*$, i.e., the constituent phases are embedded into an effective medium with yet unknown m^*, a self-consistent effective medium approximation (SCA) is captured.

For a 1–3 type composite with piezoelectric rods aligned in the magnetic matrix (denoting as 1–3 p/m), the NSCA of Eq. 4.9 gives

$$\alpha_{33}^* = -\frac{f(1-f)q_{31}e_{31}}{{}^m k + {}^m m + f({}^p k - {}^m k)}. \qquad (4.10a)$$

Inversely, for a 1–3 type composite with magnetic rods aligned in the piezoelectric matrix (denoting as 1–3 m/p), the NSCA of Eq. 4.9 becomes

$$\alpha_{33}^{*} = -\frac{f(1-f)q_{31}e_{31}}{^{m}k + ^{p}m + f(^{p}k - ^{m}k)},$$ (4.10b)

For these two kinds 1–3 microstructures, the SCA of Eq. 4.9 gives

$$\alpha_{33}^{*} = -\frac{f(1-f)q_{31}e_{31}}{^{m}k + m^{*} + f(^{p}k - ^{m}k)}.$$ (4.11)

Similarly, the constitutive Eq. 4.3 for the ME composites can also be solved by using micromechanics methods. Among them, Li [12] and Huang [13] gave more details about these micromechanics simulations. The micromechanics models are also formally straightforward and universal. It has been already known that micromechanics methods give almost the same approximations as the NSCA from Green's function technique. Figure 4.2 shows such a comparison between two approaches.

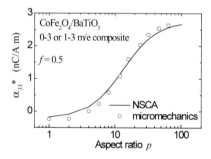

Figure 4.2 Comparison of calculated α_{33}^{*} for 0–3 or 1–3 m/p $CoFe_2O_4$/ $BaTiO_3$ ceramic composites ($BaTiO_3$ being as the matrix phase) by the micromechanical approximation and NSCA.

For the extreme case of a 1–3 fiber composites with infinite aspect ratio, Benveniste [14] and Chen [15] proposed a set of relationships between the effective properties including the effective ME coefficients by generalizing Hill's method [16] for the purely elastic case of such a fiber composite. They paid attention to the internal consistency of the solutions for the constitutive coefficient tensors.

For a 2–2 type composite (Fig. 4.1) with the piezoelectric phase poled along x_3 axis, the composite has also a ∞*mm* symmetry with ∞

denoting the x_3 axis (i.e., out-of-plane mode). In the limit case that the aspect ratio p of the layers approaches infinite, Eq. 4.8a gives similar expression for the longitudinal ME coefficient along the symmetric x_3 axis to that derived from an averaging method [17, 18].

4.1.3 Some Results

Although it is hard to get simple complicit expressions for α^* from the general solution (8) for other connectivity schemes, the general expression (4.8) is easily programmed for numerical calculations of α^* of the composites. Figure 4.3 further shows a numerical example for the ME response of the 0–3 or 1–3 m/p $CoFe_2O_4/BaTiO_3$ ceramic composites.

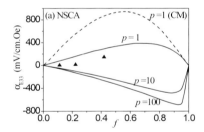

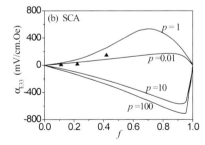

Figure 4.3 Calculated α_{E33} for 0–3 or 1–3 m/p $CoFe_2O_4/BaTiO_3$ ceramic composites ($BaTiO_3$ being as the matrix phase). The results from the simple cube model (CM) and three experimental data are also shown for comparison.

Here the ME voltage coefficient α_{E33}, describing the ME_H output voltage (on open circuit condition) developed cross the composites along the x_3 axis, is used, i.e.,

$$\alpha_{E33} = \alpha_{33}^* / \varepsilon_{33}^* \ (= E_3 / H_3)$$

(4.12)

which is the figure of merit used to assess the performance of a ME material for a magnetic device. A few points can be drawn from Fig. 4.3, i.e., (1) the simple cube model [19] overestimates the ME effect of the composites (Fig. 4.3a); (2) NSCA and SCA predict similar results for this system (Fig. 4.3a,b); (3) For 0–3 particulate ceramic composites, the ME voltage coefficient α_{E33} reaches a maximum in the middle concentration region around $f \sim 0.6$, but for 1–3 composites the maximum α_{E33} appear at $f \sim 0.9$. Thus, remarkable ME effect could be achieved in the composites with high concentrations of particulate magnetic phase well dispersed in the piezoelectric phase.

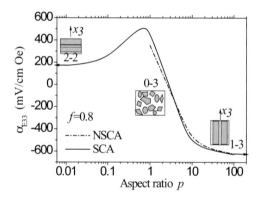

Figure 4.4 Effect of the grain shape and phase connectivity on the ME response of m/p $CoFe_2O_4$ (denoted by grey areas; volume fraction $f = 0.8$)/ $BaTiO_3$ (denoted by white areas) ceramic composites.

It is obvious that the grain shape and phase connectivity have a significant effect on the ME response of the composites. Let us still consider the case as above that the piezoelectric phase is poled along x_3 axis and the magnetic field is also along the symmetric x_3 axis. Figure 4.4 shows such an example. From 1–3 to 0–3, α_{E33} of this composite system changes its sign around $p = 3$. Due to strong anisotropy, the 1–3 structure has maximum α_{E33} along the symmetric x_3 axis, but the present case (i.e., both the magnetic field and electrical poling being along the symmetric x_3 axis) is not optimal one for the 2–2 structure. For example, if the magnetic field and/or electrical poling are perpendicular to the symmetric axis, then we can get more larger ME effect in the 2–2 structure (i.e., α_{E31} or $\alpha_{E11} > \alpha_{E33}$), as shown in Refs. [17, 18] where also was given the same results for the extreme 1–3 ($p \to \infty$) and 2–2 ($p \to 0$) composites. The authors gave explicit expressions for the ME coefficients in the two extreme cases, and especially for the 2–2 ($p \to 0$) composite with the magnetic and electrical poling perpendicular to the symmetric axis, the calculated α_{E11} can reach over 1 V/cm Oe, which is higher than that calculated for the 0–3 particulate ceramic composites.

Table 4.1 Various constituent phases used for ME ceramic composites.

Piezoelectric phase	Magnetic phase
BaTiO$_3$ (BTO)	Ni-ferrites (e.g., NFO)
PZT	Co-ferrites (e.g., CFO)
Pb(Mg,Nb)O$_3$ (PMN)	Li-ferrites (LFO)
PbTiO$_3$ (PTO)	Cu-ferrite, Mn-ferrite
(Sr,Ba)Nb$_2$O$_5$	Ytttrium iron garnet (YIG)
	(La,M)MnO$_3$ (M = Ca,Sr)

Although the calculations above focus on m/p ferrites/piezoelectric ceramic composites with the piezoelectric phase as the matrix, the insufficiently high resistivity of the ferrite phase would make it hard to obtain high ME response in the 1–3 composites and 0–3 particulate composites with high concentration of ferrite grains (i.e., large f), as expected, due to their large leakage. In comparison, the 2–2 laminate ceramic composites have no such limitation along x_3 because the ferrite layers are separated by the piezoelectric layers [2, 3]. On the other hand, it is easy to prepare the 2–2 laminate ceramic composites experimentally, while such 1–3 ceramic composites are very hard to prepare. The modeling of the 2–2 laminate ceramic composites has been discussed in details in other chapters.

4.2 TWO-PHASE COMPOSITES OF ALLOYS AND PIEZOELECTRIC MATERIALS

Rare-earth-iron alloys (e.g., $SmFe_2$, $TbFe_2$, or Terfenol-D) are the best known and most widely used giant magnetostrictive alloys exhibiting much higher magnetostriction (over 10^3 ppm) than the magnetic oxide (e.g., ferrite) ceramics. Thus, the composites of these alloys and piezoelectric materials should have much large ME response. By generalizing Green's function technique to treat the composites containing these alloys (e.g., Terfenol-D), Nan *et al.* [7] calculated the ME response of such composites, and predicted their giant ME (GME) effect.

For Terfenol-D-based composites, more generally, by considering the coupling interaction between magnetostriction (a *nonlinear* magnetomechanical effect) and piezoelectricity, the coupling response can be described by the following modified constitutive equation as [7]

$$\sigma = cS - e^T E - cS^{ms}$$
$$D = eS + \varepsilon E + \alpha H \qquad (4.13)$$
$$B = \mu(S,E,H)H,$$

where the permeability μ strongly depends on **S** and electric and magnetic fields; and S^{ms} is the magnetostrictively induced strain related with the magnetic-field-dependent magnetostriction constants, λ_{100} and λ_{111}, of the magnetostrictive alloy. By using the

Green's function technique as above, the effective ME coefficient of the composites can be solved as [7]

$$\alpha^* < \mathbf{H} > fSf\mathbf{e}^* \left\langle [\mathbf{I} - \mathbf{G}^u(\mathbf{c} - \mathbf{c}^o)]^{-1} \mathbf{G}^u \mathbf{c} \mathbf{S}^{ms} \right\rangle_{\text{Orient}}, \qquad (4.14)$$

where f is still the volume fraction of the magnetic phase; $\mathbf{c}$ and $\mathbf{c}^o$ are, respectively, the stiffness tensors of the magnetic phase and the homogeneous reference medium; $< >_{\text{Orient}}$ denotes averaging over all possible orientations of the magnetic phase in the composites. Under the open circuit measurement condition (**<D>** = 0) and a completely mechanically clamped boundary condition (**<S>** = 0), the primary ME output voltage can be obtained from equations above as

$$\bar{\mathbf{E}} = -(\kappa^*)^{-1} \alpha^* < \mathbf{H} >= \alpha_E < \mathbf{H} >= -f \, \mathbf{h}^* \left\langle [\mathbf{I} - \mathbf{G}^u(\mathbf{c} - \mathbf{c}^o)]^{-1} \mathbf{G}^u \mathbf{c} \mathbf{S}^{ms} \right\rangle_{\text{Orient}},$$

$$(4.15a)$$

where $\mathbf{h}^* (=\kappa^{*-1}\mathbf{e}^*)$ is the effective piezoelectric stress coefficient tensor. To the other extreme, there is a mechanical free boundary condition, i.e., **<σ>** = 0, and in this case, the ME output voltage is

$$\bar{E}^{\text{free}} = \bar{E} - k^{x-1}e^x \bar{S}^{ms}. \qquad (4.15b)$$

The first term in the right side of Eq. 4.15b corresponds to the primary ME response, and the second term is the secondary ME response measuring the additional ME response produced by the magnetostrictive strain of the composite through the piezoelectric effect of the composite. These equations are the desired formula for the extrinsic linear ME_H coefficient resulting from the coupling interaction between the *nonlinear magnetostrictive* and *linear piezoelectric* effects. The equations also show that a strong ME_H response of the composite could be achieved with the large magnetostrictively induced strain, the high piezoelectricity, and the perfect coupling between the phases (i.e., transferring elastic strains without appreciable losses) [9].

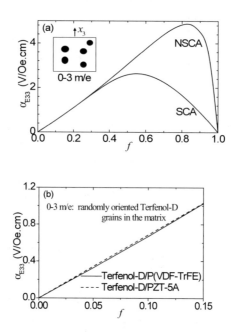

Figure 4.5 (a) Calculated α_{E33} for 0–3 m/e Terfenol-D/P(VDF–TrFE) particulate composites by NSCA and SCA. (b) Comparison of calculated α_{E33} for the flexible polymer-matrix composite and brittle PZT ceramic matrix composite filled with randomly oriented Terfenol-D particles.

For simple 0–3 particulate composites, the low resistive Terfenol-D grains should be well dispersed in the piezoelectric matrix to keep the composite insulating, since the conductive Terfenol-D grain percolation path can make it difficult to polarize the composites and cause the charges developed in the piezoelectric phase to leak through this conductive path. Thus, the Terfenol-D grains must be not in contact with each other while the piezoelectric matrix is self-connected thus forming a 0–3 connectivity of phases, and the volume fraction of the alloy grains in the piezoelectric matrix is limited by the percolation. For illustrative purpose, Fig. 4.5 shows the numerical results in the high-magnetic-field saturation in the whole volume fraction range for the 0–3 particulate Terfenol-D/P(VDF–TrFE) composite with Terfenol-D grains randomly oriented in the polymer matrix. The NSCA and SCA predict different maximum

α_{E33} at different volume fractions (e.g., at $f \sim 0.85$ and 0.55 for NSCA and SCA, respectively) and different results at high concentration, but they give very close results in the volume fraction range of $f < 0.4$, which means that the theoretical predictions in this volume fraction range of interest are reasonable. The ME coefficients of the particulate composites increase with increasing f in the volume fraction range of $f < 0.4$ of interest. The comparison between the flexible Terfenol-D/P(VDF–TrFE) composite and brittle Terfenol-D/PZT composite (Fig. 4.5b) shows that both systems exhibits similar α_{E33} due to their comparable magnetostriction and piezoelectric stress coefficient h_{33}.

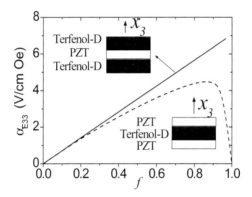

Figure 4.6 α_{E33} at high-field saturation of these two sandwiched composites as a function of the volume fraction (or relative thickness ratio) of the Terfenol-D layers.

Similarly to the ferrites/piezoelectric ceramic composites discussed above, although the 0–3 type piezoelectric-matrix composites with Terfenol-D grains embedded is simple, the 2–2 laminate composites are more realizable, since high concentration of conductive Terfenol-D can be easily separated by the piezoelectric layers in the 2–2 composites. Figure 4.6 shows numerical results for the two 2–2 sandwich composites with perfectly interfacial bonding between PZT and Terfenol-D disks. For the sandwich PZT/Terfenol-D/PZT composite, the ME effect non-monotonically depends on f with a maximum at $f \sim 0.85$, while the ME effect of the sandwich Terfenol-D/PZT/Terfenol-D composite nearly linearly increases with f. Of particularly interesting to note is that a GME effect is produced in these two laminate composites with thick Terfenol-D layers but thin PZT layers.

For the 2–2 laminate composites, Dong *et al.* [20] proposed an equivalent-circuit approach which is more convenient for modeling the ME coupling in the dynamic cases. This approach is also based on magnetostrictive and piezoelectric constitutive equations, where the magnetostrictive and piezoelectric layers are mutually coupled through elastic interaction, via an equation of motion that is excited by a magnetic field $\bar{H}$.

Table 4.2 Various constituent phases used for ME laminate composites.

Piezoelectric phase	Magnetic phase
PZT	Terfenol-D
PMN-PT, PZN-PT	Ni, Permendur, Fe-Ga, Ni_2MnGa
PVDF	Metglas

The GME effect predicted was reported first in the Terfenol-D/ PZT laminate by Ryu *et al.* [21] and in the Terfenol-D/PVDF laminate by Mori and Wuttig [22]. Since then, in particular, Dong *et al.* [8, 20, 23] have reported the GME effect in a number of laminate composites of Terfenol-D (or Metglas) and various piezoelectric materials including PZT ceramics, $Pb(Mg_{1/3}Nb_{2/3}O_3)$–$PbTiO_3$ (PMN-PT) or $Pb(Zn_{1/3}Nb_{2/3}O)_3$–$PbTiO_3$ (PZN-PT) single crystal, or electroactive PVDF co-polymers. ME laminates can be made in many different configurations including disc, rectangular, and ring shapes. These various configurations can be operated in numerous working modes including T-T (transverse magnetization and transverse polarization), L-T, L-L, symmetric L-L (push–pull) longitudinal vibrations; L-T unimorph and bimorph bending; T-T radial and thickness vibrations multilayer; and C-C (circumferential magnetization and circumferential polarization) vibration modes.

The laminate composites are generally fabricated by bonding magnetostrictive and piezoelectric layers (Table 4.2) using an epoxy resin, followed by annealing at a modest temperature of 80–100°C. In the calculations above, a perfectly bonded interface between magnetic alloys and piezoelectric phases is assumed, which ensures the perfect coupling between these two phases (i.e., transferring elastic strains without appreciable losses) to obtain exceptional control of the predictability of the ME effect in such ME composites.

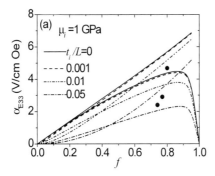

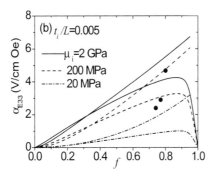

Figure 4.7 Effect of (a) relative thickness t_i/L (L being the thickness of the composite) and (b) shear modulus μ_i of the interfacial binder layers on α_{E33} at high-field saturation of these sandwich composites. Recent experimental data available for a Terfenol-D/PZT/Terfenol-D composite are also shown for comparison.

Any imperfect interfaces will more or less decrease the displacement transfer capability, thereby leading to a decrease in the ME response of the composites. To treat the interfacial bonding effect, we change the shear modulus μ_i and the relative thickness t_i of the interfacial binder [24]. Figure 4.7 shows the influence of interface properties on α_{E33} of these sandwich composites. The comparison in Fig. 4.7 shows that the predictions are in reasonable agreement with experimental data available, though the interfacial bonding status in the experiment is not clear. The increase in the thickness t_i of the interfacial binder films between the PZT and Terfenol-D leads to a decrease in the α_{E33} values (Fig. 4.7a). Figure 4.7a shows that a very thin binder film (e.g., $ti/L \sim 0.001$) with bonding capability is good enough for producing the GME response. The elastic modulus of the interfacial binder films has a significant effect on the α_{E33} values (Fig. 4.7b). With decreasing the shear modulus μ_i of the thin binder films (i.e., using a very flexible binder), the interfacial bonding between the PZT and Terfenol-D becomes weak due to the formation of a sliding interface. The weak interfacial contact would lead to appreciable losses of transferring elastic strain/stress from Terfenol-D to PZT, and thus the decrease in the shear modulus μ_i of the thin interfacial binder films results in a large decrease in the ME response of the composites.

4.3 THREE-PHASE COMPOSITES

In the two-phase Terfenol-D-based laminates, the Eddy current loss in bulk Terfenol-D is quite large at high frequency, which limits the applications of this laminate structure, and the GME response of the laminated composites is strongly influenced by the interfacial binders [24], as discussed above. In addition, Terfenol-D thin layers are very brittle. To overcome these difficulties in two-phase Terfenol-D/PZT laminate composites, a three-phase composite of Terfenol-D, PZT, and a polymer has been developed [9]. Such three-phase composites can be easily fabricated by a conventional low-temperature processing, and especially the three-phase particulate composites can be fabricated into a variety of forms such as thin sheets and molded shapes, and they exhibit greatly improved mechanical properties.

The simplest three-phase ME composite is quasi 0–3 type particulate composites where Terfenol-D grains are randomly oriented in the mixture matrix of PZT and polymer [9]. The conductive Terfenol-D grains are well dispersed in the PZT/polymer matrix to keep the composite insulating, as schematized in Fig. 4.8a, thus forming a 0–3–3 connectivity of phases. In the three-phase ME composites, the inactive polymer is just used as a binder. Such a three-phase particulate composite shown in Fig. 4.8 can be denoted as f Terfenol-D/$(1 - f_m - f)$PZT/f_m polymer, where f and f_m are the volume fractions of Terfenol-D and polymer, respectively. The ME response of the 0–3–3 composites can also be calculated by using the Green's function technique as discussed above [25]. The solid line in Fig. 4.8b shows calculated α_{E33} (under the boundary condition of completely mechanically clamped in the x_3 direction but free in the transverse direction as in experimental measurement) of the f Terfenol-D/$(0.7 - f)$PZT/0.3polymer composite. As shown, in the low volume fraction range of $f < 0.1$, the α_{E33} values at high-magnetic-field saturation increase approximately linearly with increasing f. This increase in α_{E33} is attributed to the increase in magnetostrictively induced strain of the composites with f. Comparison in Fig. 4.8b shows that there is a good agreement between the experimental and calculated and measured α_{E33} as $f < 0.07$. However, as $f > 0.07$, the measured α_{E33} values decline with f, which is due to formation of a Terfenol-D grain percolation path through the composites, though the magnetostrictively induced strain of the composites still increases with f. The maximum α_{E33} value of the composites at 2 kOe is about 42 mV/cmOe at $f = 0.06$, which is comparable with those measured for the ferrite/PZT ceramic particulate composites.

As observed, the ME effect in the Terfenol-D/PZT/PVDF particulate composites is mainly limited by the concentration threshold of the Terfenol-D grains allowed in the composites due to the low resistance of Terfenol-D, which is dependent on the composite microstructure. If f allowed in the composites is higher, the α_{E33} values will increase. The results imply that the ME response of the three-phase composites could be further improved by carefully controlling processing to achieve homogeneity and higher loading of the Terfenol-D grains well-isolated in the PZT/polymer mixture.

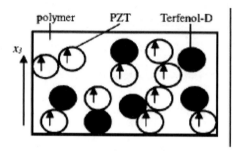

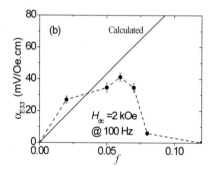

Figure 4.8 (a) Schematic representation of the particulate Terfenol-D/PZT/ polymer composites. All grains are randomly embedded in the polymer matrix. The polarization in PZT particles (denoted by the arrows within the open circles) is assumed to be parallel to the x_3 direction. (b) Calculated (solid line) and measured (dots) α_{E33} at high field saturation for the f Terfenol-D/ $(0.7 - f)$PZT/0.3polymer composites.

To eliminate this limitation on the low volume fraction of Terfenol-D in the quasi 0–3 type particulate composites, 2–2 type laminate composites containing Terfenol-D/polymer and PZT/ polymer layers have been made by laminating the Terfenol-D/ polymer (denoted as T-layer) and the PZT/polymer (denoted as P-layer) particulate composite layers and then simply hot-molding these layers together [3]. Such a simple hot-pressing procedure ensures a good interfacial bonding between composite layers, since the polymer used in the T-layer and P-layer is the same just as a matrix binder. Different laminate structures, can be obtained, e.g., sandwich structure of P-layer/T-layer/ P-layer (abbreviated as the P-T-P composite), or T-layer/ P-layer/T-layer (abbreviated as the T-P-T composite), or transverse structure of P-layer and T-layer. In comparison with the quasi 0–3 type particulate composites, these quasi 2–2 type laminate composites exhibit higher the ME effect and strong anisotropy, and thus different longitudinal ME coefficient α_{E33} and transverse ME coefficient α_{E31}. The ME response of the three-phase polymer-based laminate composites could be further improved by optimization of their structures, and processing, and would be an important smart multiferroic material for magnetic–electric devices.

Besides 0–3 and 2–2 structures, the 1–3 structure, i.e., a fiber (or rod) reinforced composite, is another important structure. As for the 1–3 piezoelectric composites, such 1–3 ME composite [26] can be prepared via a dice-and-fill process. According to the designed volume fraction of PZT rods, a PZT bulk is diced to get a PZT rod array, and then the gap of the PZT rod array is filled with a mixture of Terfenol-D particles and epoxy. When the epoxy hardens, the pseudo 1–3 type multiferroic composite is obtained (see the inset of Fig. 4.9).

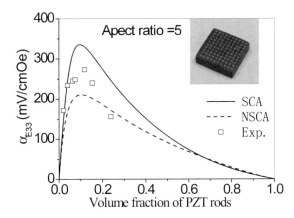

Figure 4.9 Variation in the ME coefficient of the pseudo 1–3 type multiferroic composites (see the inset) with the volume fraction of PZT rods at low frequency. Solid and dashed lines in (a) are calculated results by SCA and NSCA.

The dependence of the ME coefficients on the volume fraction f_{PZT} of PZT rods is presented in Fig. 4.9. The calculations and experiments all show that the ME coefficient increases with increasing f_{PZT} due to the increasing piezoelectric effect. After reaching a maximum, the ME coefficient decreases when $f_{PZT} > 0.1$. This decrease is attributed to the decrease in the volume fraction of the magnetostrictive Terfenol-D/Epoxy matrix. The maximum α_{E33} reaches about 300 mV/cm·Oe at 2 kOe, which is comparable with that for the pseudo 2–2 laminate ME composites and larger than that for the 0–3 particulate ME composites.

4.4 NANOSTRUCTURED COMPOSITE THIN FILMS

Following Zheng *et al.* [10], ME nanocomposite thin films of ferroelectric [e.g., BTO, PTO, PZT, and BiFeO$_3$ (BFO)] and magnetic oxides (e.g., CFO, NFO, and LCMO) prepared by physical deposition techniques (e.g., pulsed laser deposition) and chemical solution processing (e.g., sol–gel spin-coating method) have recently become new routes to multiferroic ME composites [27]. In comparison to bulk ME composites, the nanostructured thin films provide more degrees of freedom, such as lattice strain or interlayer interaction, for modifying the ME behavior. These films also offer a way to investigate the physical mechanism of the ME effect in nanoscale. The coupling interaction between two oxides in the multiferroic nanostructures is still due to an elastic interaction as was the case in bulk composites. However, the mechanical constraints arising from the film–substrate bonding and the bonding between the two phases in the nanostructured composite films could significantly affect the ME coupling interactions.

Similarly to bulk ceramic composites shown in Fig. 4.1, defining on the microstructure of the nanostructured composite films, there are also three kinds of nanostructured composite films, i.e., (1) 0–3 type structures with magnetic spinel nanoparticles (e.g., CFO and NFO) embedded in the ferroelectric films (e.g., PZT), (2) 1–3 type heterostructures (vertical heterostructures) consisting of magnetic spinel pillars (e.g., CFO) vertically embedded into a ferroelectric films (e.g., BTO, PTO, or BFO), and (3) 2–2 type heterostructures (horizontal nanostructures) consisting of alternating layers of a ferroelectric perovskite and magnetic oxide.

In theoretical analyses, some characteristics of films, e.g., giant residual stress/strain resulting from the lattice misfit between the film and substrate, and spontaneous polarization in epitaxial films, have been considered to understand the ME response in nanostructured films [28]. As a result, the constitutive equations for the coupling magnetic–mechanical–electric interactions in the nanostructured films can be expressed as [28]

$$\sigma = c\varepsilon - e^T E - c\varepsilon^{ms} - \sigma_s ,$$
$$D = e\varepsilon + \kappa E + \alpha H + P_s , \qquad\qquad (4.16)$$
$$B = \mu(\varepsilon, E, H)H + M_s .$$

In comparison to the constitutive equations for bulk composites, the residual stress σ_s (or residual strain ε_s), spontaneous polarization P_s, and magnetization M_s are incorporated for the multiferroic composite films. Based on the modified constitutive equations, the ME coupling interaction and magnetically induced polarization for nanostructured composite films were firstly calculated using the Green's function technique [28].

$$\bar{P} = (1-f)(P_S + (k^p - k^*)G^\phi (P_S - e^p G^u \sigma_S^p)$$
$$+ (e^* - ep)G^u (\sigma_S^p + e^T G^\phi P_S)) \qquad\qquad (4.17)$$
$$+ fe^* (I - G^u (c^m - c^p))^{-1}(c^m \varepsilon^{ms} + \sigma_S^m),$$

where f is the volume fraction of the ferromagnetic phase, and the quantities denoted using the superscripts p and m refer to those for the ferroelectric and ferromagnetic phases.

For the 1–3 type nanostructured composite films with $(00l)$ ferromagnetic nanopillars embedded in a ferroelectric matrix, the effective electric polarization $\bar{P}_3$ along the symmetric direction is easily obtained from Eq. 4.17 as [28]

$$\bar{P}_3 = (1-f)P_{S3} + \frac{2f(1-f)e_{31}^p}{\bar{k} + c_{11}^p - c_{12}^p}(\sigma_{11}^p - \sigma_{11}^m) \qquad\qquad (4.18)$$

in the case that the height of pillars is much larger than their diameter (e.g., 20–30 nm in diameter and 400 nm in height for nanopillars [10]), where σ_{11}^p and σ_{11}^m are in-plane total stresses including residual stresses and magnetostrictively induced stresses, respectively, in the ferroelectric and ferromagnetic phases.

For the 2–2 type nanostructured composite films with (001) ferroelectric and ferromagnetic nanolaminates, the effective electric polarization $\overline{P}_3$ can also be directly obtained from Eq. 4.17 as [28]

$$\overline{P}_3 = (1-f)\frac{k_{33}^m}{k_{33}}P_{S3} + \frac{f(1-f)e_{31}^p k_{33}^m}{\overline{k}_{33}\overline{c}_{33}}(\sigma_{33}^p - \sigma_{33}^m) \qquad (4.19)$$

in the case that the thickness of a layer (e.g., 30 nm [10]) is much less than the macroscopic size of the film plane, where σ_{33}^p and σ_{33}^m are out-of-plane total stresses including residual stresses and magnetostrictively induced stresses, respectively, in the ferroelectric and ferromagnetic phases.

These explicit equations show that the effective electric polarization is sensitive to residual strains and spontaneous polarization in the films, materials constants of the two phases and the ways of their combination, and applied magnetic field. It was revealed that the 1–3 type vertical heterostructures could exhibit large ME response which is even larger than that in their bulk counterparts if there is no leakage problem, while the 2–2 type horizontal heterostructures show much weaker ME coupling due to large in-plane constraint effect. Most recently, the calculations for the ME effects in nanobilayers, nanopillars, and nanowires of nickel ferrite and PZT on MgO substrates or templates have also shown that the ME coupling decreases with increasing substrate clamping [29].

Because the magnetostriction of the magnetic phase can dynamically change the strain constraint in the nanostructured composite films, the magnetically induced polarization could also be calculated from the Landau–Ginsburg–Devonshire phenomenological thermodynamic theory [30], where the boundary conditions are related to the magnetic field. This method was found to yield some similar results to those using the Green's function technique, i.e., the calculation results showed that the 1–3 type nanostructured composite films exhibit a large ME coefficient, but the 2–2 type films exhibit much weaker ME coefficient due to large in-plane constraint.

A powerful approach, phase-field model, in which the elastic energy in the constrained thin film was incorporated including the effect of free film surface and the constraint from the substrate, has been recently developed for studying the ME coupling effect in the 1–3 nanocomposite thin films [31]. The phase-field calculations of the magnetic-field-induced electric polarization in 1–3 type $BaTiO_3$–$CoFe_2O_4$ nanocomposite film give similar results to the Green's function technique. The phase-field simulation illustrates that the magnetic-field-induced electric polarization is highly dependent on the film thickness, morphology of the nanocomposite, and substrate constraint, which provide a number of degrees of freedom in controlling coupling in nanocomposite films.

ME nanostructures have become an important topic of ever-increasing interest in last few years, since they, especially ME thin films, are easy to on-chip integration, which is a prerequisite for incorporation into microelectronic devices. But ME nanostructures are just in infant stage. Many open questions regarding the ME coupling in nanostructures remain.

4.5 CONCLUSIONS

In bulk ME composites including ceramic composites, two-phase piezoelectric–magnetic alloy composites, and three-phase polymer-based composites, the ME coupling effect is a strain-mediated one which can be well simulated by the physics-based approach. The effective coupling properties of these ME composite have been expressed in a convenient matrix formulation by using the successful Green's function technique. This approach not only described the influence of the composite microstructure (e.g., phase volume fraction, phase connectivity, aspect ratio of the phase particles) on the effective properties of ME composites, but also predicted the GME effect in composites containing Terfenol-D. In the ME nanostructures such as nanocomposite thin films, the ME coupling is still strain-mediated in most cases. Such strain-mediated ME coupling in the

ME nanostructures can also be described by the Green's function technique. The physics-based approach for the ME composites remains to be further generalized to treat the dynamic ME behavior of these ME composites including bulk and nanostructures.

Acknowledgments

Many excellent papers on this topic are not cited due to the limited space and I sincerely apologize for that. I thank my research associates and graduate students, and collaborators (specially, Professor Mirza Bichurin's group in Novgorod State University, Russia), for their contributions. This work was supported by Ministry of Science and Technology of China and the National Science Foundation of China.

References

1. J. Van Suchtelen, "Product properties: a new application of composite materials," Philips Res. Rep., **27**, 28 (1972).

2. M.I. Bichurin, D. Viehland, G. Srinivasan, "Magnetoelectric interactions in ferromagnetic-piezoelectric layered structures: phenomena and devices," J. Electroceram., **19**, 243 (2007).

3. C.W. Nan, M.I. Bichurin, S. Dong, D. Viehland, G. Srinivasan, "Multiferroic magnetoelectric composites: Historical perspective, status and future directions," J. Appl. Phys., **103**, 031101 (2008).

4. C.W. Nan, "Magnetoelectric effects in composites of piezoelectric and piezomagnetic phases," Phys. Rev., **B50**, 6082 (1994).

5. J. Van den Boomgard *et al.* "An in situ grown eutectic magnetoelectric composite materials: part I," J. Mater. Sci., **9**, 1705 (1974).

6. G. Srinivasan *et al.* "Magnetoelectric bilayer and multilayer structures of magnetostrictive and piezoelectric oxides," Phys. Rev. B **64**, 214408 (2001).

7. C.W. Nan, M. Li, J.H. Huang, "Calculations of giant magnetoelectric effects in ferroic composites of rare-earth-iron alloys and ferroelectric polymers," Phys. Rev., **B63**, 144415 (2001).

8. S. Dong, J.F. Li, D. Viehland, "Ultrahigh magnetic field sensitivity in laminate of TERFENOL-D and $Pb(Mg_{1/3}Nb_{2/3})O_3$-$PbTiO_3$ crystals," Appl. Phys. Lett., **83**, 2265 (2003).

9. C.W. Nan, L. Liu, N. Cai, J. Zhai, Y. Ye, Y.H. Lin, L.J. Dong, C.X. Xiong, "A three-phase magnetoelectric composite of piezoelectric ceramics, rare-earth iron alloys, and polymer," Appl. Phys. Lett., **81**, 3831 (2002).

10. H. Zheng *et al*, "Multiferroic $BaTiO_3$-$CoFe_2O_4$ nanostructures," Science, **303**, 661 (2004).

11. C.W. Nan, D.R. Clarke, "Effective properties of ferroelectric and/or ferromagnetic composites: a unified approach and its application," J. Am. Ceram. Soc., **80**, 1333 (1997).

12. J.Y. Li, M.L. Dunn, "Anisotropic coupled-field inclusion and inhomogeneity problems," Philos. Mag., **A77**, 1341 (1998).

13. J.H. Huang, "Analytical predictions for the magnetoelectric coupling in piezomagnetic materials reinforced by piezoelectric ellipsoidal inclusions," Phys. Rev., **B58**, 12 (1998).

14. Y. Benveniste, "Magnetoelectric effect in fibrous composites with piezoelectric and piezomagnetic phases," Phys. Rev., **B51**, 16424 (1995).

15. T.Y. Chen, "Exact moduli and bounds of two-phase composites with coupled multifield linear responses," J. Mech. Phys. Solids, **45**, 385 (1997).

16. R. Hill, "Theory of mechanical properties of fibre-strengthened materials: I. elastic behaviour," J. Mech. Phys. Solids, **12**, 199 (1964).

17. I. Getman, "Magnetoelectric composite materials: theoretical approach to determine their properties," Ferroelectrics, **162**, 45 (1994).

18. M.I. Bichurin, V.M. Petrov, G. Srinivasan, "Theory of low-frequency magnetoelectric coupling in magnetostrictive-piezoelectric bilayers," Phys. Rev., **B68**, 054402 (2003).

19. G. Harshe, J.P. Dougherty, R.E. Newnham, "Theoretical modelling of multilayer magnetoelectric composites," Int. J. Appl. Electromagn. Mater. **4**, 145 (1993).

20. S. Dong, J.F. Li, D. Viehland, "Magneto-electric coupling, efficiency and voltage gain effect in piezoelectric-piezomagnetic laminate composite: theory and analysis," J. Mater. Science, **41**, 97 (2006).

21. J. Ryu, S. Priya, A.V. Carazo, K. Uchino, H.E. Kim, "Effect of magnetostrictive layer on magnetoelectric properties in lead

zirconate titanate/terfenol-D laminate composites," J. Am. Ceram. Soc., **84**, 2905 (2001).

22. K. Mori, M. Wuttig, "Magnetoelectric coupling in terfenol-D/ polyvinylidenedifluoride composites," Appl. Phys. Lett., **81**, 100 (2002).

23. S. Dong, J.F. Li, D. Viehland, "Characterization of magnetoelectric laminate composites operated in longitudinal-transverse and transverse-transverse modes," J. Appl. Phys., **95**, 2625 (2004).

24. C.W. Nan, G. Liu, Y.H. Lin, "Influence of interfacial bonding on giant magnetoelectric response of multiferroic laminated composites of $Tb_{1-x}Dy_xFe_2$ and $PbZr_xTi_{1-x}O_3$," Appl. Phys. Lett., **83**, 4366 (2003).

25. Z. Shi, C.W. Nan, J. M. Liu, M.I. Bichurin, "Influence of mechanical boundary conditions and microstructural features on magnetoelectric behavior in a three-phase multiferroic particulate composite," Phys. Rev., **B70**, 134417 (2004).

26. Z. Shi, C.W. Nan, J. Zhang, N. Cai, J.F. Li, "Magnetoelectric effect of $Pb(Zr,Ti)O_3$ rod arrays in a $(Tb,Dy)Fe_2$/epoxy medium," Appl. Phys. Lett., **87**, 012503 (2005).

27. R. Ramesh, N.A. Spaldin. "Multiferroics: progress and prospects in thin films," Nature Mater., **6**, 21 (2007).

28. C.W. Nan, G. Liu, Y.H. Lin, H. Chen, "Magnetic-field-induced electric polarization in multiferroic nanostructures," Phys. Rev. Lett., **94**, 197203 (2005).

29. V.M. Petrov, G. Srinivasan, M.I. Bichurin, A. Gupta, "Theory of magnetoelectric effects in ferrite piezoelectric nanocomposites," Phys. Rev., **B75**, 224407 (2007).

30. G. Liu, C.-W. Nan, J. Sun, "Coupling interaction in nanostructured piezoelectric/magnetostrictive multiferroic complex films," Acta Mater., **54**, 917 (2006).

31. J.X. Zhang, Y.L. Li, D.G. Schlom, L.Q. Chen, F. Zavaliche, R. Ramesh, Q.X. Jia, "Phase-field model for epitaxial ferroelectric and magnetic nanocomposite thin films," Appl. Phys. Lett., **90**, 052909 (2007).

Chapter 5

Equivalent Circuit Method and Magnetoelectric Low-Frequency Devices

S. Dong[a] and D. Viehland[b]

[a]Department of Advanced Materials and Nanotechnology,
College of Engineering, Peking University,
Beijing 100871, China
[b]Department of Materials Science and Engineering, Virginia Tech,
Blacksburg, VA 24061, USA

The magneto–elasto–electric equivalent circuits for L–L, L–T, and T-T modes were developed. The proposed equivalent circuit method is quite useful for both static and dynamic analysis of magnetoelectric (ME) laminates, especially for electromechanical resonance analysis. The equivalent circuit method predicts that ME coupling in ME laminates is strongly related to their working modes: the value of ME voltage coefficients in L–T or L–L modes is significantly higher relative to the T-T. Another critical predictions are that there is an optimum thickness ratio of piezoelectric and magnetostrictive layers that maximizes ME coefficients and resonance ME voltage coefficient is $\sim Q_{\mathrm{m}}$ times higher under resonance drive. These predictions have been confirmed by experimental results.

Magnetoelectricity in Composites
Edited by Mirza I. Bichurin and Dwight Viehland
Copyright © 2012 Pan Stanford Publishing Pte. Ltd.
www.panstanford.com

5.1 EQUIVALENT CIRCUIT METHOD: THEORY

Theoretical analysis of the magnetoelectric (ME) effect for magnetostrictive/piezoelectric two-phase composites has been performed using constitutive equations and a composite averaging method [5, 6, 15]. These approaches simply modify the piezoelectric constitutive equation by including the corresponding magnetostrictive one. The ME coupling in laminates has also been modeled by a Green's functional approach [12], and in the case of fiber composites, by an eigenstrain formulation [16]. For the 2–2 or 2–1 laminate composites, Dong *et al.* [9, 14, 17, 18] proposed an equivalent circuit approach which is more powerful for modeling ME coupling in the dynamic cases. This approach is also based on magnetostrictive and piezoelectric constitutive equations, where the magnetostrictive and piezoelectric layers are mutually coupled through elastic interaction (strain $S(z)$ and stress $T(z)$), via an equation of motion that is excited by a magnetic field $\vec{H}$.

5.1.1 Three-Layer L–T and L–L Longitudinal Vibration Modes

We suppose that the magnetostrictive/piezoelectric laminates are long-type configurations, in which the piezoelectric layer is polarized along either its thickness or length directions, and stressed by two magnetostrictive layers along their length (i.e., principal strain) direction. Accordingly, the piezoelectric constitutive equations for one-dimensional motion are

$$S_{1p} = s_{11}^{E} T_{1p} + d_{31,p} E_3 \qquad D_3 = d_{31,p} T_{1p} + \varepsilon_{33}^{T} E_3 , \tag{5.1a}$$

for thickness poling, and

$$S_{3p} = s_{33}^{D} T_{3p} + g_{33p} D_3 \qquad E_3 = -g_{33p} T_{3p} + \beta_{33}^{T} D_3 , \tag{5.1.b}$$

for length polarization case, where D_3 is the electric displacement, ε_{33}^{T} and β_{33}^{T} are the dielectric permittivity and impermeability under constant stress T, s_{11}^{E} and s_{33}^{D} are the elastic compliances of the piezoelectric material under constant electric field E or constant electric displacement D, $d_{31,p}$ and $g_{33,p}$ are the transverse piezoelectric constant and longitudinal piezoelectric voltage constants, and T_{1p}, T_{3p} and S_{1p} S_{3p} are the stress and strain of the piezoelectric layer imposed by the magnetostrictive layers.

When H is applied parallel to the longitudinal axis of the laminate, a longitudinal (33) strain is excited. The piezomagnetic constitutive equations for the (33) longitudinal mode are

$$S_{3m} = s_{33}^H T_{3m} + d_{33,m} H_3 \qquad\qquad B_3 = d_{33,m} T_{3m} + \mu_{33}^T H_3 , \qquad (5.2)$$

where B_3 is the magnetization along the length direction, μ_{33}^T is the permeability under constant stress, s_{33}^H is the elastic compliance of the magnetostrictive layer under constant H, $d_{33,m}$ is the longitudinal magnetostrictive constant, and T_{3m} and S_{3m} are the stress and strain in the longitudinal direction of the magnetostrictive layers imposed on the piezoelectric layer. These constitutive equations are linear relationships, which do not account for loss components. Significant nonlinearities in both piezoelectric and magnetostrictive materials are known to exist, especially under resonance drive. We will introduce a mechanical quality factor Q_m to include these losses later.

Assuming harmonic motion, along a given direction z, it will be supposed that three small mass units, Δm_i, in the laminate have the same displacement $u(z)$. This follows from Fig. 5.1, by assuming that the layers in the laminate act only in a coupled manner. Following Newton's Second Law, we then have an equation of motion to couple the piezoelectric and piezomagnetic constitutive Eqs. 5.1 and 5.2 as

$$\bar{\rho}\frac{\partial^2 u(z)}{\partial t^2} = n\frac{\partial T_{3,m}}{\partial z} + (1-n)\frac{\partial T_{i,p}}{\partial z} \qquad (0 < n < 1,\ i = 1\ \text{or}\ 3), \qquad (5.3)$$

where $\bar{\rho} = \dfrac{\rho_p A_p + \rho_m A_m}{A_{lam}}$ is the average mass density of the laminate, and $n = A_m/A_{lam} = t_m/t_{lam}$ is a geometric factor, $t_m = t_{m1} + t_{m2}$ is the total thickness of the magnetic phase layers, A_p and A_m ($= A_{m1} + A_{m2}$) are the cross-sectional areas of the magnetic phase and piezoelectric phase layers respectively, ρ_p and ρ_m are the mass densities of the piezoelectric and magnetostrictive layers. For a given laminate width w_{lam} and thickness t_{lam}, the total cross-sectional area of the laminate is $A_{lam} = t_{lam} w_{lam}$, and the total thickness is the sum of the layer thicknesses $t_{lam} = t_p + t_m$, where t_p is the thickness of the piezoelectric layer.

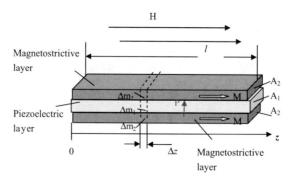

Figure 5.1 Long type magnetostrictive/piezoelectric/magnetostrictive laminate.

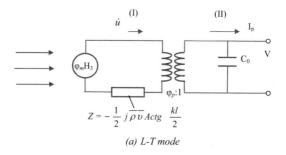

(a) L-T mode

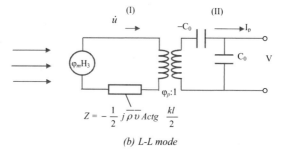

(b) L-L mode

Figure 5.2. Magneto-elastic-electric bi-effect equivalent circuits for (a) L–T mode, where $\varphi_m = 2A_2 \dfrac{d_{33,m}}{s_{33}^H}$, $\varphi_p = \dfrac{wd_{31,p}}{s_{11}^E}$; $C_0 = \dfrac{lw}{t_1}\varepsilon_{33}^T$; and (b) L–L mode,

where $\varphi_m = 2A_2 \dfrac{d_{33,m}}{s_{33}^H}$; $\varphi_p = \dfrac{A_1 g_{33p}}{ls_{33}^D \beta_{33}}$; $C_0 = \dfrac{A_1}{l\beta_{33}}$. See also Color Insert.

By combining piezoelectric and piezomagnetic constitutive Eqs. 5.1 and 5.2, solutions to the equation of motion (5.3) can be derived for longitudinally magnetized M and transversely or longitudinally poled P ME modes, named as L–T or L–L. Correspondingly, the two magneto–(elasto)–electric (or ME) equivalent circuits for L–T and L–L modes under free boundary conditions can be derived, as given in Figs. 5.2a,b [9, 17]. In these figures, an applied magnetic field H acts as a magnetic induced "mechanical voltage" ($\varphi_m H_3$), and then induces a "mechanical current" $\dot{u}$ via the magnetoelastic effect with a coupling factor φ_m. In turn, $\varphi_m H_3$ results in an electrical voltage V, and $\dot{u}$ results in a real electric current I_p across the piezoelectric layer, due to electromechanical coupling. A transformer with a turn-ratio of φ_p can then be used to represent the electromechanical coupling. In the circuits of Fig. 5.2, Z_1 and Z_2 are the characteristic mechanical impedances of the composite, and C_0 is the clamped capacitance of the piezoelectric plate.

5.1.2 ME Voltage Coefficients at Low Frequency [9, 14, 17, 18]

Following Fig. 5.2a,b, open-circuit conditions have been supposed from above, where the current I_p from the piezoelectric layer is zero. Thus, the capacitive load C_0 (and $-C_0$) can be moved to the main circuit loop. Applying Ohm's law to the mechanical loop, the following ME field coefficients at low-frequency for L–T and L–L modes can be directly derived as

$$\left| \frac{dE}{dH_3} \right|_{(L-T)} = \frac{nd_{33,m}g_{31,p}}{ns_{11}^E(1-k_{31,p}^2)+(1-n)s_{33}^H} , \tag{5.4a}$$

$$\left| \frac{dE}{dH_3} \right|_{L-L} = \frac{nd_{33,m}g_{33,p}}{ns_{33}^E(1-k_{33}^2)+(1-n)s_{33}^H} . \tag{5.4b}$$

These are the formulas for the ME field coefficients predicted by the equivalent method. The values $\left|\dfrac{dE}{dH_3}\right|$ of are proportional to the longitudinal piezomagnetic constant $d_{33,m}$, and to the transverse or longitudinal piezoelectric voltage constants $g_{31,p}$ or $g_{33,p}$. In addition, the ME voltage coefficient is strongly related to the thickness ratio n of the Terfenol-d layers, and to the elastic compliances of both phases. Also, please note that these predicted ME electric field coefficients from Eqs. 5.4a,b are much higher than measured the values in our prior reports [9, 14, 17, 18]; therefore, a rectifying factor β was used to modify these formulas.

5.1.3 ME Coefficients at Resonance Frequency [14, 18]

Assuming long-type ME laminate composites to be a L–L $\lambda/2$-resonator, operating in a length extensional mode, the series angular resonance frequency is $\omega_s = \dfrac{\pi \bar{v}}{l}$, where l is the length of the laminate, and $\bar{v}$ is mean acoustic velocity of the laminate composite. Under resonant drive, the mechanical quality factor Q_m of the laminate is finite due to both mechanical and electric dissipations. This limitation of the vibration amplitude must also be included, in order to predict the resonant response. Finite values of Q_{mech} result in an effective motional mechanical resistance of $R_{mech} = \dfrac{\pi Z_0}{8 Q_{mech}}$ accordingly, the equivalent circuit of the laminates for the L–L mode under resonance drive is given in Fig. 5.3. At electromechanical resonance ($\omega = \omega_s$), dV/dH of the L–L mode reaches a maximum value of

$$\left(\frac{dV}{dH}\right)_{És} = \frac{4 Q_m \varphi_m \varphi_p}{\pi Z_0 \omega_s C_0} , \tag{5.5}$$

where Q_m is the effective mechanical quality factor of the laminate composite including contributions from the Terfenol-D and piezoelectric layers, and also from the bonding between layers.

Analysis has shown that dV/dH at the resonance frequency is $\sim Q_{\mathrm{m}}$ times higher than that at sub-resonant frequencies. Using a similar approach, it is easy to obtain the resonance equivalent circuit for the L–T mode, but is not shown here.

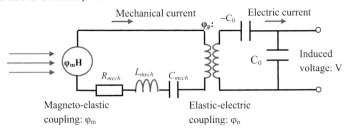

Figure 5.3 Magneto-elastic-electric equivalent circuits for L–L at resonance [14,18], where $L_{\mathrm{mech}} = \dfrac{\pi Z_0}{8\omega_s}$, $C_{\mathrm{mech}} = \dfrac{1}{\omega_s^2 L_{\mathrm{mech}}}$, and $Z_0 = \overline{\rho \upsilon} A_{\mathrm{lam}}$, $R = \dfrac{\pi Z_0}{8Q_m}$. See also Color Insert.

5.1.4 Two-Layer L–T Bending Mode [42]

The Terfenol-D/PZT two-layer laminate, as shown in Fig. 5.4, is an unimorph operated in L–T bending mode, where magnetostrictive layer is magnetized in length (longitudinal) direction and the piezoelectric layer is poled in thickness (transverse) directions. Under an applied ac magnetic field H_{ac}, the magnetostrictive layer in Terfenol-D/PZT laminates strain along the longitudinal-axis, exciting a bending motion mode due to an unsymmetric strain about the longitudinal neutral-face (line). Using piezoelectric and piezomagnetic constitutive equations within a small bending elastic theory, the ME field coefficient under a low-frequency (quasi-static) drive can be derived as

$$\frac{dE}{dH} \approx \frac{d_{33,m} g_{31,p} t_{\mathrm{m}} (t_{\mathrm{m}} - 2t_{\mathrm{n}})(t_{\mathrm{p}} + 2t_{\mathrm{m}} - 2t_{\mathrm{n}})}{4 S_{33}^H (1 - v_{\mathrm{p}}) D_{\mathrm{comp}} / Y_{\mathrm{p}}}, \tag{5.6}$$

where $d_{33,m}$ is the longitudinal piezomagnetic coefficient for the longitudinal magnetization mode; $g_{31,p}$ the transverse piezoelectric voltage coefficients; t_{m} the thickness of the magnetostrictive layer; t_{p} the thickness of the piezoelectric layers; t_{n} the neutral-face position of the Terfenol-D/PZT composite; s_{33}^H the elastic compliances of the magnetostrictive layer, Y_p and v_p Young's Modulus and Poisson's Ratio of the piezoelectric layers; and D_{comp} the composite stiffness of the laminate. The flexural rigidity (stiffness) of the composite is a resultant of that of the constituent layers given as

$$D_{comp} = D_m + D_p - t_m (t_m - 2t_n)^2 k_{33,m}^2 / 4S_{33}^B,$$ (5.7)

where $k_{33,m}$ is the magnetoelastic coupling coefficient, and D_m and D_p the bending stiffness of the magnetostrictive and piezoelectric layers, respectively. From Eq. 5.6, we can see that dE/dH is proportional to (i) $d_{33,m}$ $(\delta\lambda_{33}/\delta H)$ of the magnetostrictive layer and (ii) $g_{31,p}$ of the piezoelectric layer, but dE/dH is reverse proportional to D_{comp}. Clearly, a ME laminate with low stiffness will be favorable to high ME coupling.

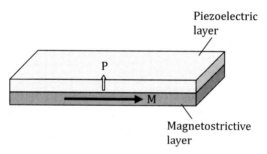

Figure 5.4 L–T bending mode.

5.1.5 Three-Layer C–C Radial Vibration Mode

In the C–C mode, because the piezomagnetic and piezoelectric layers are mutually coupled via strain $S(z)$ and stress $T(z)$, application of an ac vortex magnetic field H_{ac} along the circumferential direction of the magnetostrictive rings puts the piezoelectric one into forced oscillation along the same direction. This excites a radial symmetric vibration mode in the piezoelectric ring, generating a voltage across each segment of the piezoelectric ring, via the longitudinal piezoelectric constant $d_{33,p}$. Supposing that this laminated ring is thin, the transverse (31) vibration mode (width or thickness) can be neglected, relative to its circumferential principle strain. Accordingly, the voltage induced by the magnetic field is a C–C mode effect. We suppose that each segment in the piezoelectric ring is polarized along the circumferential direction, accordingly, the piezoelectric constitutive equations are

$$S_{3p} = s_{33}^E T_{3p} + d_{33,p} E_3 \qquad\qquad D_3 = d_{33,p} T_{3p} + \varepsilon_{33}^T E_3. \qquad (5.8)$$

Because H_{ac} is applied parallel to the circumferential direction of the magnetostrictive rings, a longitudinal (33) strain will be excited. Therefore, we should use the piezomagnetic constitutive equations (Eq. 5.2) to express this (33) longitudinal mode.

We suppose that the circumferential principle vibration imposed on the Terfenol-D layers induces a symmetric radial vibration in the laminated ring. For a given laminated ring of width w_{lam} and thickness t_{lam}, the total cross-sectional area of the laminate (A_{lam}) is the sum of the magnetostrictive (A_m) and piezoelectric (A_p) layer areas $(A_{lam} = A_p + 2A_m = t_{lam} \omega_{lam})$, and the total thickness is the sum of the magnetostrictive (t_m) and piezoelectric (t_p) layer thicknesses $(t_{lam} = t_p + 2t_m)$. From Fig. 5.5, the resultant force F_r imposed on an incremental segment $\delta\theta$ of the laminated ring is then

$$F_r = F \times \delta\theta = (2F_m + F_p) \times \delta\theta, \qquad (5.9a)$$

where F_m and F_p are the forces on the respective layers. This, along with Newton's second law, yields

$$w(2t_m \rho_m + t_p \rho_p) a \delta\theta \frac{\partial^2 \xi}{\partial t^2} = -w(2t_m T_{3m} + t_p T_{3p}) \delta\theta, \qquad (5.9b)$$

where ρ_m and ρ_p are the densities of the Terfenol-D and PZT rings, respectively; and ξ and a are the radial displacement and mean radius of the ring, respectively. (Note that $S_{3p} = S_{3m} \approx \xi/a$.)

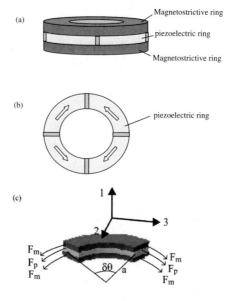

Figure 5.5 ME C–C laminate.

Now we can use this equation of motion to couple the piezoelectric Eq. 5.8 and piezomagnetic Eq. 5.2, and use Eq. 5.8 to find the induced current I and voltage V. Then, the following relationships can be found [35] as

$$\dot{\xi} = \frac{\varphi_m H_3}{Z_m} + \frac{\varphi_p V}{Z_m},$$ (5.10a)

$$I = \varphi_p \dot{\xi} + j\omega m C_0 V,$$ (5.10b)

$$Z_m = j\omega L_m + \frac{1}{j\omega C_m} + R_m,$$ (5.10c)

where $\varphi_m = \dfrac{4\pi A_m d_{33,m}}{s_{33}^H}$ is the magnetoelastic coupling factor; $\varphi_p = \dfrac{m A_p d_{33,p}}{a s_{33}^E}$ the elasto–electri coupling factor; $C_0 = (1 - k_{33p}^2)C^T$

and $C^T = \dfrac{A_p \varepsilon_{33}^T}{2\pi a / m}$ are the clamped and free capacitances of a single piezoelectric segment; Z_m the mechanical impedance, containing mechanical inductance $L_m = 2\pi a A_{\text{lam}} \bar{\rho}$ ($\bar{\rho} = n\rho_m + (1-n)\rho_p$ and $n = 2A_m/A_{\text{lam}}$), mechanical compliance as $C_m = \dfrac{1}{L_m \omega_0^2}$, and mechanical loss in the laminate (an effective motional mechanical resistance) as $R_m = \omega_0 L_m / Q_m$, where ω_0 ($\omega_0 = \bar{v}/a$ and $\bar{v}^2 (\dfrac{n}{s_{33}^H} + \dfrac{1-n}{s_{33}^E})/\bar{\rho})$ is the free angular resonance frequency of the laminated ring, and Q_m is the mechanical quality factor of the laminate.

A combined magneto–(elasto)–electric equivalent circuit for the C–C mode (assuming no external load) can be developed, as given in Fig. 5.6, where the symbol (I) is used to designate the magnetic section, (II) the mechanical section, and (III) the electric section.

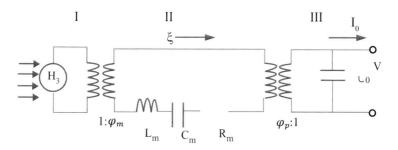

Figure 5.6 Equivalent circuit of ME C–C mode. See also Color Insert.

Expressions for the ME voltage coefficients were derived from the equivalent circuit of Fig. 5.6, as follows. Under open-circuit conditions, the current I from the piezoelectric layer is zero. Thus, the capacitive load mC_0 can be moved to the mechanical portion of the circuit. Following Ohm's law in Fig. 5.6, the ME voltage coefficient under a small magnetic field excitation can be directly derived as

$$\left| \frac{dV}{dH_3} \right| = \left| \frac{\varphi_p \varphi_m}{\varphi_p^2 + j\omega m C_0 Z_m} \right| . \qquad (5.11a)$$

For $\omega \ll \omega_0$, the mechanical losses can be neglected. Thus, the value of the mechanical impedance is $Z_m \sim \dfrac{1}{j\omega C_m}$. Then, ME voltage coefficient can be approximated as

$$\left|\frac{dV}{dH_3}\right| = \frac{\varphi_p\varphi_m}{\dfrac{mC_0}{C_m} + \varphi_p^2}. \tag{5.11b}$$

Equation 5.11b shows that the ME voltage coefficient is dependent upon φ_m of the Terfenol-D rings and φ_p of the PZT one. Clearly, high magnetoelastic and electromechanical couplings are important contributing factors to the design of a high ME voltage coefficient in laminated rings. Equation 5.11b can be expressed as dE/dH, and we can confirm that it has a completely same expression as that of an L–L mode, as shown in Eq. 5.4b.

At electromechanical resonance, $\omega = \omega_0$, and $j\omega L_m + \dfrac{1}{j\omega C_m} = 0$. From the equivalent circuit, it is easy to find ME voltage coefficient at resonance:

$$\left|\frac{dV}{dH_3}\right| = \left|\frac{\varphi_p\varphi_m}{\varphi_p^2 + j\omega_0^2 mC_0 L_m / Q_m}\right| \tag{5.11c}$$

Again, we can see that dV/dH at the resonance frequency is $\sim Q_m$ times higher than that at sub-resonant frequencies.

5.1.6 Analysis on ME Voltage Gain [14, 45, 46]

As we previously mentioned, the ME voltage coefficient of dV/dH at the electromechanical resonance is $\sim Q_m$ times higher than that at sub-resonant frequencies. This strong ME coupling at resonance can be used to produce voltage gain, see Fig. 5.7. When a solenoid with N turns around a ME laminate that carries a current of I_{in}, it will excite an ac magnetic field H_{ac}, and causes the two magnetostrictive layers to shrink/expand in response to H_{ac}. The magnetostrictive strain acts upon the piezoelectric layer that is bonded between the two magnetostrictive layers, causing the piezoelectric layer to strain, producing a voltage output. An input ac voltage applied to the coils is V_{in}, and its frequency is f. This excites a H_{ac} of the same frequency f, along the longitudinal direction of the laminate.

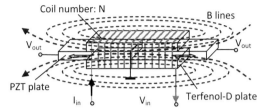

(a) Configuration of ME transformer.

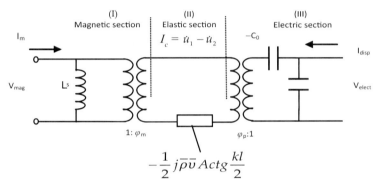

(b) Equivalent circuit at low frequency.

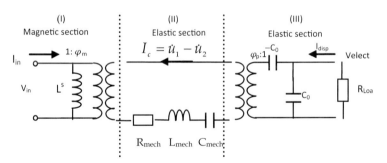

(c) Equivalent circuit at resonance.

Figure 5.7 ME transformer. See also Color Insert.

When the frequency of H_{ac} is equal to the resonance frequency of the laminate, the ME coupling effect is so strong that the output ME voltage V_{out} induced in the piezoelectric layer is much higher than V_{in}. Thus, under resonant drive, there is a high voltage gain, due to the resonant ME effect.

To obtain a maximum ME voltage gain, the polarization direction of the piezoelectric layer was chosen to be along its length direction under longitudinal vibration, as shown in Fig. 5.7a. This is because both the piezoelectric constant g_{33} and electromechanical coupling coefficient k_{33} in the longitudinal mode are 2× that of the transverse g_{31} and k_{31} coefficients. Consequently, higher output voltages V_{out} can be obtained from the ME laminate.

Application of H along the length direction of the laminate excites a longitudinal $(d_{33,m})$ mode in the magnetostrictive layer. Two sets of constitutive linearized equations are required to describe the coupled responses of the piezoelectric and magnetostrictive layers. These are piezoelectric constitutive Eq. 5.1b and piezomagnetism constitutive equation:

$$T_{3m} = \frac{1}{s_{33}^B} S_{3m} - \lambda_{33} B_3 , \qquad H_3 = -\lambda_{33} S_{3m} + v_{33}^s B_3$$

$$v_{33}^s = \frac{1}{\mu_{33}^s} , \quad \mu_{33}^s = \mu_{33}^T (1 - k_{33,m}^2), \quad s_{33}^B = s_{33}^H (1 - k_{33,m}^2),$$

$$k_{33,m}^2 = \frac{d_{33,m}^2}{s_{33}^H \mu_{33}^T} , \quad \lambda_{33} = \frac{d_{33,m}}{s_{33}^H \mu_{33}^s} \tag{5.12}$$

where λ_{33} is the magnetostrictive coefficient; v_{33}^s is the magnetic stiffness (reluctivity) under constant strain; μ_{33}^s and μ_{33}^T are the magnetic permeabilities under consta nt strain and stress.

Upon considering the insulating layer in the laminate, we have to introduce a third constitutive equation:

$$\varepsilon_{3I} = \frac{1}{Y^I} \sigma_{3I} , \tag{5.13}$$

where ε_{3I}, σ_{3I}, and Y^I are longitudinal strain, stresses and Young's modulus of the insulating layer, respectively.

Under harmonic motion along $\hat{z}$ (longitudinal direction), it can be supposed that all small mass units Δm_i in the magnetostrictive, piezoelectric, and insulation layers of the laminate at z have the same displacement $u(z)$ or strain ε_z without any sliding between layers. Following Newton's Second Law, we then have

$$\sum_i \Delta m_i \frac{\partial^2 u}{\partial t^2} = \sum_i \sigma_{3i} A_i , \tag{5.14}$$

where i denotes the layer number, and A_i cross-section area of ith layer. We can then use this motion Eq. 5.14 to couple the three-phase constitutive Eqs. 5.1b, 5.12, and 5.13. By using these equations, following magneto–elasto–electric relationships can be found as [14]

$$F_1 = Z_1 \dot{u}_1 + \left(Z_2 + \frac{\varphi_p^2}{j\omega(-2C_0)} \right)(\dot{u}_1 - \dot{u}_2) + \varphi_p V_{out} + \varphi_m V_{in},$$

$$F_2 = -Z_1 \dot{u}_2 + \left(Z_2 + \frac{\varphi_p^2}{j\omega(-2C_0)} \right)(\dot{u}_1 - \dot{u}_2) + \varphi_p V_{out} + \varphi_m V_{in},$$

$$(5.15)$$

$$I_{in} = \frac{\lambda_{33}}{j\omega N}(\dot{u}_1 - \dot{u}_2) + \frac{V_{in}}{j\omega L^S} \qquad (5.16)$$

where $Z_1 = j\bar{\rho}\bar{v}A_{lam} \cdot tg\left(\frac{kl}{2} \right)$, $\quad Z_2 = \frac{\bar{\rho}\bar{v}A_{lam}}{j\sin(kl)}$, $\quad \varphi_m = \frac{\lambda_{33}}{j\omega N}$,

$$\varphi_p = \frac{A_p g_{33,p}}{s_{33}^D \bar{\beta}_{33} l/2}, \quad C_0 = \frac{A_p}{\bar{\beta}_{33} l/2}, \quad \bar{\beta}_{33} = \beta_{33}^T \left(1 + \frac{g_{33,p}^2}{s_{33}^D \beta_{33}^T} \right),$$

$$Ls = Am\mu_{33}^s N2/l, \quad k^2 = \frac{\omega^2}{\bar{v}^2}, \quad \bar{v}^2 = \left(\frac{n_m}{s_{33}^B} + \frac{n_p}{s_{33}^D} + n_1 \gamma' \right)/\bar{\rho},$$

$$n_m = \frac{A_m}{A_{lam}}, \quad n_p = \frac{A_p}{A_{lam}}, \quad n_l = \frac{A_l}{A_{lam}}. \qquad (5.17)$$

Equations 5.15 and 5.16 completely describe the magneto(–elastic–)electric coupling between the magnetostrictive and piezoelectric layers under an exciting current I_{in} or voltage V_{in}, via the mechanical displacement velocities $\dot{u}_1$ and $\dot{u}_2$. According to impedance electromechanical analogies, the equivalent circuit of ME transformer under free boundary conditions can be derived as shown in Fig. 5.7b. In this circuit, the input current I_{in} (or voltage V_{in}) excites a "mechanical current" $I_c = \dot{u}_1 - \dot{u}_2$, via the magnetoelastic coupling factor φ_m. Subsequently, I_c induces an output voltage V_{out}, via the elasto–electric coupling factor φ_p.

Assuming the laminate composite to be a length extensional $\lambda/2$-resonator, the series angular resonance frequency is $\omega_s = \frac{\pi\bar{v}}{l}$. Under resonant drive, the mechanical character impedance Z in the equivalent circuit of Fig. 5.2b can be approximated by a Taylor series expansion of the frequency $f(\omega)$ about ω_s. After comparing with the impedance expansion of a series L_m, C_m circuit and including a mechanical dissipation impedance Z_m, equivalent circuit of the laminate under resonance drive is finally given in Fig. 5.7c,

where $\quad L_{mech} = \dfrac{\pi Z_0}{8\omega_s}; \quad C_{mech} = \dfrac{1}{\omega_s^2 L_{mech}}; \quad R_{mech} = \dfrac{\omega_s L_{mech}}{Q_{mech}} = \dfrac{\pi Z_0}{8 Q_{mech}};$

$\omega_s^2 = \dfrac{1}{L_{mech} C_{mech}} \quad Z_0 = \overline{\rho \upsilon A_{lam}}$, and R_{Load} is an external load.

According to equivalent circuit Fig. 5.7c [14], some important relationships, such as effective ME coupling factor, optimum thickness ratio, maximum voltage gain, and energy conversion efficiency etc., can be found as follows.

5.1.6.1 Effective ME coupling factor

$k_{mag-elec}^2 \left(eff \right) =$

$$\dfrac{64 s_{33}^B k_{33,m}^2 g_{33p}^2 n_m n_p}{\pi^2 \overline{\beta}_{33} (n_m s_{33}^D + n_p s_{33}^B + n_l Y^I s_{33}^B s_{33}^D)[\pi^2 / 2(n_m s_{33}^D + n_p s_{33}^B + n_l Y^I s_{33}^B s_{33}^D) + 8 n_m k_{33m}^2 s_{33}^D]},$$

where $s_{33}^B = s_{33}^H (1 - k_{33,m}^2)$, $s_{33}^D = s_{33}^E (1 - k_{33,p}^2)$. $\qquad\qquad$ (5.18)

Optimum geometric parameter for the magnetostrictive layer (assuming the insulating layer to be thin):

$$n_{m,opt} = \dfrac{1}{1 + \gamma(1 + 8 k_{33m}^2 / \pi^2)^{1/2}}, \qquad\qquad (5.19)$$

where $\gamma = \dfrac{s_{33}^D}{s_{33}^B}$ is a ratio of the compliance constants of the piezoelectric and magnetostrictive layers.

Maximum ME voltage gain under resonance drive
i. Directly derived voltage gain:

$$V_{gain1,max} = \dfrac{4 Q_{mech} \varphi_m \varphi_p}{\pi \omega_s C_0 Z_0}. \qquad\qquad (5.20)$$

ii. Voltage gain after introducing a ratio factor:

$$V_{\text{gain2,max}} = \frac{4Q_{\text{mech}}\varphi_{\text{p}}^2}{\pi\omega_s C_0 Z_0} . \tag{5.21}$$

Calculation results show that Eq. 5.21 gives more close values to measured ones.

5.1.6.2 Maximum efficiency

$$\eta_{\text{max}} = \frac{\varphi_{\text{p}}^2}{\varphi_{\text{p}}^2 + \dfrac{\pi Z_0 C_0 \omega_s}{Q_{\text{mech}}}} \quad \text{when} \quad R_{\text{Load,opt}} = \frac{1}{2\omega_s C_0} . \tag{5.22}$$

Clearly, a higher Q_{mech} will results in higher transduction efficiency. For $Q_m = 100$, η_{max} was less than 90%; however, for $Q_{\text{mech}} = 1000$, η_{max} was $\approx 98\%$. However, for a bulk Terfenol-D material operated of a high frequency of >20 kHz, the Eddy current losses will be serious, resulting in that the effective ME efficiency far lower than the theoretical value. To overcome the Eddy current losses, a thin multi-layer design is necessary for high-power ME transformer applications.

5.1.6.3 Analysis on ME gyration

Gyration from current-to-voltage (I–V) is important for device applications in power conversions [49, 50], high-sensitivity current sensors, current-output digital-to-analog converters [51], filters and communication devices [52] etc. Conventional I–V gyrations are based on active operational amplifiers and switching converters [53–55], and passive (sourceless) Faraday Effect in ferrites [56, 57]. In these I–V gyrators or converters, the gyration coefficients are low, and the gyrated output voltages are not high. A simple resistance element is also capable of I–V conversion (not gyration). Unfortunately, it is a power-consumed element, not suited for power conversion applications. Fortunately, we have found that ME laminates showed gyration effect, which has potential as passive I–V converters [47].

We consider a long-type ME laminate [45], consisting of one strong piezoelectric, high-Q_m (mechanical quality factor) PbZr$_x$Ti$_{1-x}$O$_3$ (PZT) layer that is symmetrically poled (**P**) along in length direction sandwiched between two strong magnetostrictive $Tb_{1-x}Dy_xFe_2$ (Terfenol-D) layers magnetized (**M**) in their length (longitudinal) direction. This laminate is illustrated in Fig. 5.8, and is designated as the "push–pull" configuration. A longitudinally applied magnetic field H will cause the Terfenol-D layers magnetized and produce an elastic strain longitudinally via magnetostrictive effect, which is then coupled to the laminated piezoelectric PZT layer stressing it along the longitudinal direction. This will result in a polarization change Δ**P** in piezoelectric layer (piezoelectric effect), generating net free-charges $q+$ at the two end electrodes and $2q-$ at middle electrode of the piezoelectric layer. Figure 5.8 illustrates this magneto-to-electric conversion (ME effect). If H is an ac magnetic field, this will excite a symmetric expanding/contracting elastic strain or vibration about the center line of the laminate, and an ac electric voltage is produced across the end and middle electrodes of the piezoelectric layer. Suppose that this ac magnetic field, $H_{ac} = \dfrac{NI_{in}}{l}$, is excited by a solenoid wrapped around the laminate, where I_{in} is the input ac current, N the coils turn number, and l the length of the laminate. The input current I_{in} then induces an output voltage (V_{out}) via the ME effect, realizing passive current-to-voltage or I–V conversion. Under electromechanical resonant drive, this passive I–V conversion will reach its maximum. Apparently, this I–V conversion is a bi-coupling effect, consisting of magnetoelastic and elastoelectric effects. Because this is a strong piezoelectric/strong magnetostrictive two-phase laminated composite, it will definitely show higher ME coupling than that in single-phase or multi-phase particulate systems.

An "ideal gyrator" was first reported in 1948 by Tellegen [7]. An ideal gyrator is a passive network component that acts as anti-reciprocal couple. Our recent investigations [47, 48] have shown that ME laminate composites have some characters of such Tellegen gyrators, such as a 180° phase shift between input current and output voltage, or vice versa. Theoretical analysis of ME gyration is still founded on piezoelectric and magnetostrictive constitutive equations, mutually coupled to each other through elastic strain $S(z)$ and stress $T(z)$. An equation of motion, driven by an ac I_{in} in the coils, is used to couple the two constitutive equations. Due to elastic coupling to the magnetostrictive layers, the piezoelectric layer is also put into forced oscillation, generating a voltage across its two end electrodes and middle ground plane. Accordingly, the following relationships can be derived:

$$F_1 = -Z_1\dot{u}_1 + \left(Z_2 + \frac{\varphi_p^2}{j\omega(-2C_o)}\right)(\dot{u}_1 - \dot{u}_2) + G\varphi_p I_{in} + \varphi_p V_{out} \quad (5.23a)$$

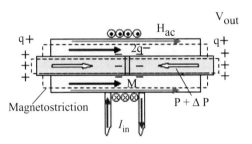

(a) ME gyration configuration.

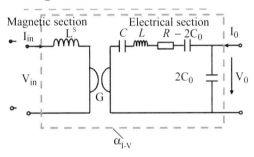

(b) Equivalent circuit of ME gyrator equivalent circuit model of ME gyrator under resonance drive and free-condition, where $G = \varphi_m / \varphi_p$;

$$\varphi_m = \frac{NA_m d_{33,m}}{s_{33}^H l}; \quad \varphi_p = \frac{2A_p g_{33,p}}{l s_{33}^D \overline{\beta}_{33}}; \quad L^S = \frac{\mu^s N^2 A_m}{l}; \quad R = \frac{\pi Z_0}{8Q_m \varphi_p^2}; \quad L = \frac{\pi Z_0}{8\omega_s \varphi_p^2};$$

$$C = \frac{\varphi_p^2}{\omega_s^2 L_{mech}}; \quad C_0 = \frac{2A_p}{l\overline{\beta}_{33}}; \quad Z_0 = \overline{\rho}\overline{v}A_{lam}; \quad \text{and} \quad \omega_s = \frac{\pi\overline{v}}{l}. \quad \text{The parameters}$$

A_m, A_p, and A_{lam} are the cross-sectional areas of the magnetostrictive layers, piezoelectric layer, and laminate, respectively; l is the length of the laminate; $\overline{\rho}$ and $\overline{v}$ are the mean density and acoustic velocity of the laminate; μ^s is the magnetic permeability of the magnetostrictive layer under constant stress; Q_m is the mechanical quality factor of the laminate; and N is the turn number of coils.

Figure 5.8. ME gyrator [47]. See also Color Insert.

$$F_2 = -Z_1 \dot{u}_2 + \left(Z_2 + \frac{\varphi_p^2}{j\omega(-2C_0)} \right)(\dot{u}_1 - \dot{u}_2) + G\varphi_p I_{in} + \varphi_p V_{out}, \quad (5.23b)$$

$$V_{in} = j\omega L^S I_{in} + G\varphi_p(\dot{u}_1 - \dot{u}_2), \quad (5.23c)$$

$$I_{out} = j\omega(2C_0)V_{out} + \varphi_p(\dot{u}_2 - \dot{u}_1), \quad (5.23d)$$

where F_1 and F_2 are the force phasors; $\dot{u}_1$ and $\dot{u}_2$ the mechanical velocities; V_{in} and I_{in} the input voltage and current applied to the coils; V_{out} and I_{out} the induced voltage and current from the piezoelectric layer; Z_1 and Z_2 the mechanical impedances; L^S the clamped inductance of the coil; C_0 the static capacitance of the piezoelectric layer; φ_p the electromechanical coupling factor; ω the angular frequency; and G a special coupling or transfer factor, designated as the gyration coefficient.

Next, following Tellegen [7], we introduce a passive four-terminal-network gyrator, where the voltage and current to input ends of the gyrator are I_1 and V_1, and correspondingly those to the output end are I_2 and V_2. By definition, the gyrator satisfies the following relation: $V_1 = GI_2$ and $V_2 = GI_1$. According to this gyration concept, an equivalent circuit model containing an ideal gyrator G, with negligible electric/magnetic dissipations and under free-boundary conditions, can be obtained as given in Fig. 5.8b. An applied I_1 (I_{in}) to the coils gyrates an output voltage of V_2 (V_{out}) in the piezoelectric section of the composite through a gyrator G, where G acts as an impedance inverter. Alternatively, it is possible that a current I_2 input to the piezoelectric section will also gyrate a voltage V_1 across the coils. Note that network components: L, C, R, $2C_0$ and $-2C_0$ in this equivalent circuit model act as a frequency transfer function, which effectively modulates the gyration coefficient G; and consequently, limit the I–V conversion. From Fig. 5.8b, it is straight forward to determine the I–V conversion coefficient that we designate as a_{I-V}. This is the "apparent gyration coefficient" of the "black-box" outlined by dashed lines in the figure. Assuming the output to be in an open-circuit condition, this effective parameter

a_{I-V} can be determined to be

$$\alpha_{I-V} = \frac{V_{out}}{I_{in}} = Z_R(f)G, \; G = \frac{NA_m d_{33,m} g_{33,p}}{2k_{33,p}^2 A_p s_{33}^H},$$
(5.24a)

where $Z_R(f) = \dfrac{1/j\omega(2C_0)}{R + j\omega L + 1/j\omega C}$ is a ratio of output impedance to input impedance in the electrical section of equivalent circuit model Fig. 5.8, indicating a frequency transfer function of this *I–V* converter; and *G* is the ideal gyration coefficient of the ME laminate, determined by the configuration and material parameters of the ME composite.

Clearly, a_{I-V} is a non-ideal gyration coefficient, reduced from its ideal value *G* by a factor $Z_R(f)$. At resonance, $\omega_s = \pi\bar{v}/l$ and $\dfrac{1}{j\omega C} + j\omega L = 0$; thus, the maximum current-to-voltage conversion coefficient is

$$\alpha_{I-V,max} = \frac{4\varphi_m \varphi_p Q_{m,eff}}{\pi C_0 \omega_s Z_0},$$
(5.24b)

or $V_{0,max} = \alpha_{I-V,max} I_{in}$ (at resonance),
(5.24c)

where φ_m is the magnetoelastic coupling factor. The value of $V_{0,max}$ is linearly proportional to the input current I_{in}, and a high effective mechanical quality factor, $Q_{m,eff}$, will result in large *I–V* conversion. Calculations for *G*, $Z_R(f)$, and a_{I-V} were then performed using the material parameters reported in [45]. These calculations were done by assuming a Terfenol-D/PZT laminate length, width and thickness of 70 mm, 10 mm, and 7 mm, respectively; a coils turn number of *N* = 100; and an effective mechanical quality factor of $Q_{m,eff}$ = 50. Figure 5.9 shows the predicted values for *G*, $Z_R(f)$, and a_{I-V} as a function of drive frequency *f*. It can be seen that (i) the ideal gyration coefficient *G* = 3860 V/cmOe is a constant (at a given frequency range), and is related only to the configuration of the ME laminate and to the materials parameters of its layers; (ii) the frequency transfer function $Z_R(f)$ reduces or modulates the value of the gyration coefficient from its ideal value of *G* to an effective value of a_{I-V}; and (iii) below the resonance frequency range the value of $Z_R(f)$ is small (0.02) resulting a low value of a_{I-V}, whereas at resonance it reaches a maximum value (0.64) resulting a maximum value of

a_{I-V} (2480 V/cmOe) which is not too much lower than the ideal value of G (3860 V/cmOe).

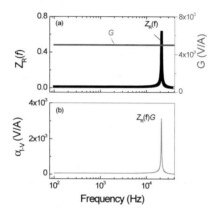

Figure 5.9 Calculations using Eq. 5.24 for (a) ideal gyration coefficient G, and frequency transfer function $Z_R(f)$; and (b) I–V conversion coefficient a_{I-V}, as a function of drive frequency f. See also Color Insert.

5.2 EXPERIMENTS

In two-phase composite systems, magnetostrictive/piezoelectric laminate layers have shown much stronger ME coupling than that in two-phase particulate composites. This is in part because large Eddy current losses restrain the effective ME coupling in particulate composites. Laminate composites are generally fabricated by bonding magnetostrictive and piezoelectric layers using an epoxy resin, followed by annealing at a modest temperature of 80–100°C. Typically, laminate composites are made of two magnetostrictive layers and a single piezoelectric one, where the piezoelectric layer is sandwiched between the two magnetostrictive ones. However, ME laminates can be made in many different configurations including disc, rectangular, and ring shapes. These varying configurations can be operated in numerous working modes including T–T (transverse magnetization and transverse polarization) [11, 19, 20], L–T, L–L, symmetric L–L (push–pull) longitudinal vibrations [8–10, 13, 17, 21–26]; L–T unimorph and bimorph bending [27–30]; T–T radial and thickness vibrations multilayer [16, 31–34]; and C–C (circumferential magnetization and circumferential polarization) vibration modes [35–37], as shown in Figs. 5.10–5.13.

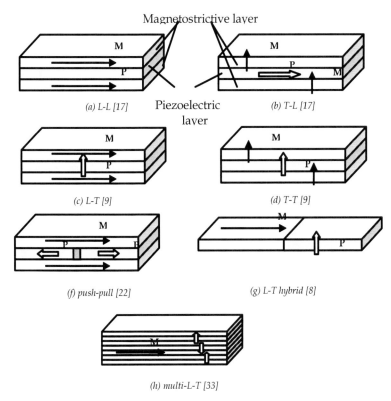

Figure 5.10 Rectangular laminates operated in longitudinal vibration mode.

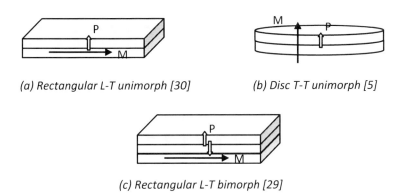

Figure 5.11 Unimorph and bimorph bending vibration mode.

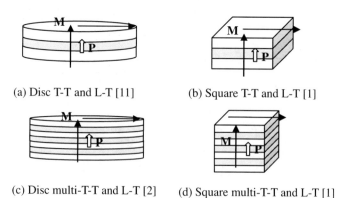

(a) Disc T-T and L-T [11]   (b) Square T-T and L-T [1]

(c) Disc multi-T-T and L-T [2]   (d) Square multi-T-T and L-T [1]

Figure 5.12 Radial and thickness vibration modes of disc and square type ME.

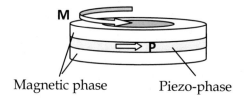

Magnetic phase   Piezo-phase

Figure 5.13 C–C mode of ring ME laminate [35].

Giant ME effects in laminate composites of super high magnetostrictive $Tb_{1-x}Dy_xFe_{2-y}$ (Terfenol-D) have been reported with various piezoelectric material couples including $Pb(Zr,Ti)O_3$ (PZT) ceramics, $Pb(Mg_{1/3}Nb_{2/3}O_3)$– $PbTiO_3$ (PMN–PT or $Pb(Zn_{1/3}Nb_{2/3}O)_3$–$PbTiO_3$ (PZN–PT) single crystal, or electroactive poly(vinylidence-fluoride) (PVDF) co-polymers [8, 9, 17–25, 27–30, 37].

5.2.1 T–T Terfenol-D/PZT Laminate

The first studies of Terfenol-D/PZT laminates [20] were performed on disc-shape three-layer configurations that were operated in a transverse magnetization and transverse polarization(T–T) mode. Relatively large ME field coefficients of a_{ME} = 4.8 V/cmOe were reported by Ryu *et al.* [20] under dc magnetic field bias of $H_{dc} \geq 4000$ Oe, although later, other investigators repeated the actual value to be $a_{ME} = 1.3$ V/cmOe [58]. Figure 5.14 shows these later (repeated) experimental results for a T–T laminate, where a maximum ME voltage of 66 mV/Oe (1.32 V/cmOe) was observed under $H_{dc} \approx 4000$ Oe.

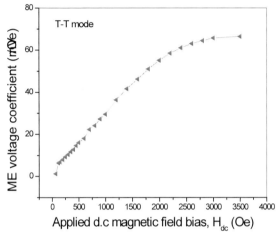

Figure 5.14 ME voltage coefficient as a function of H_{dc} [24]. See also Color Insert.

The main problem with the T–T mode laminates is the quite high dc magnetic bias H_{dc} that is required to obtain a maximum value of a_{ME}. This high H_{dc} is caused by a large demagnetization factor (N) in transversely magnetized Terfenol-D layers.

5.2.2 L–T Terfenol-D/PZT and PMN–PT Laminates

To reduce the demagnetization factor (N) effect, a long-type configuration that uses a longitudinal magnetization was designed [9, 13, 25, 26]. This dramatic decrease in N resulted in a large reduction in the H_{dc}, which is required to achieve the maximum ME coefficient [10]. Long rectangular-shaped Terfenol-D/PZT/Terfenol-D and Terfenol-D/PMN–PT/Terfenol-D three-layer laminates with a longitudinal magnetization and transverse polarization (L–T) were then reported by Dong *et al.* [9, 10, 13] based on this consideration. Experimental results confirmed at low magnetic biases of $H_{dc} < 500$ Oe that much large values of a_{ME} could be obtained for L–T laminates relative to T–T ones. Figure 5.15 shows these measurements taken under an ac magnetic excitation of $H_{ac} = 1$ Oe at $f = 1$ kHz. It can be seen that the induced ME voltage for the L–T Terfenol-D/PZT laminates under $H_{dc} = 500$ Oe was 0.085 V/Oe (or $a_{ME} = 1.7$ V/cmOe) and for L–T Terfenol-D/PMN–PT ones ~0.11 V/Oe ($a_{ME} = 2.2$ V/cmOe), whereas that for the T–T mode of Terfenol-D/PZT laminate under $H_{dc} = 500$ Oe was only ~0.015 V/Oe ($a_{ME} = 0.3$ V/cmOe). Clearly, long-type L–T laminates have significantly higher ME voltage coefficients than T–T ones under modest magnetic biases. Note that the measured voltage at $H_{dc} = 0$ is normally specified as a noise signal because the piezomagnetic coefficient d_{33m} of the Terfenol-D at $H_{dc} = 0$ is zero.

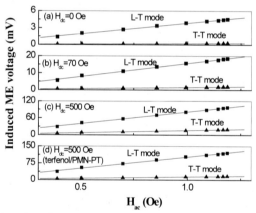

Figure 5.15 Induced magneto–electric voltage as a function of H_{ac} for a laminate of Terfenol-D and PZT-5 ceramic: (a) $H_{dc} = 0$ Oe, (b) $H_{dc} = 70$ Oe, and (c) $H_{dc} = 500$ Oe; and (d) for a laminate of Terfenol-D and a <001>-oriented PMN–PT crystal. The measurement frequency was 10^3 Hz [1].

5.2.3 L–L and Push–Pull Terfenol-D/PZT and PMN–PT Laminates

To achieve a high output voltage, a L–L mode of Terfenol-D/ PZT/Terfenol-D or Terfenol-D/PMN–PT/Terfenol-D laminate is a good choice, due to its large dielectric displacement **D** along the length (longitudinal) direction [17, 23, 24]. For a long-type ME laminate, the length of the piezoelectric layer is much larger than its thickness, in addition, the longitudinal electromechanical coupling coefficient $k_{33,p}$ and piezoelectric voltage constant $g_{33,p}$ are higher than the corresponding transverse $k_{31,p}$ and $g_{31,p}$ ones. Following Eq. 4b, a Terfenol-D/PZT laminate operated in an L–L mode should have a much higher induced voltage V_{ME} under magnetic field excitation. Figure 5.16 shows measurements for an L–L mode Terfenol-D/PZT laminate, which exhibits a maximum V_{ME} = 3.5 V/Oe at f = 1 kHz (or a_{ME} = 2.4 V/cmOe) at H_{dc} = 500 Oe. Please note that although L–L ME laminates have the highest induced voltage (in V/Oe), its ME charge coefficient (in C/Oe) is quite low. This is because L–L ME laminates have a very low capacitance. For a ME laminate with small capacitance, direct measurements of the induced ME voltages using a lock-in amplifier may result in big error [17]. In this case, a charge measurement method may be more reliable for obtaining the correct values.

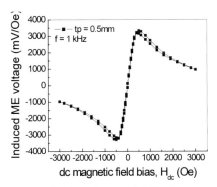

Figure 5.16 L–L Terfenol-D/PZT/Terfenol-D laminate.

A double (or symmetrically poled) L–L configuration has been designated as push–pull mode [18, 22], whose illustration was given either in Fig. 5.10f. The push–pull configuration is a compromised design, which takes advantages of the high induced voltage of L–L mode and the large ME charge of the L–T one. Figure 5.17 shows measurements of the ME voltage coefficients for a push–pull laminate, which had a maximum ME induced voltage of 1.6 V/Oe (or a_{ME} = 2.5 V/cmOe) at f = 1 kHz and ~20 V/Oe (~31 V/cmOe) at f = 75 kHz. Although this value is similar to an L–L mode, its ME charge coefficient was 2× higher than that in L–L one [22].

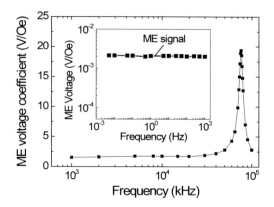

Figure 5.17 ME voltage coefficients as a function of magnetic field frequency [22].

5.2.4 L–T Bending Mode of Terfenol-D/PZT Laminates [28–30]

A single longitudinally magnetized Terfenol-D layer laminated together with one or two transversely poled piezoelectric PZT or PMN–PT layers is an L–T bending mode, as illustrated in Fig. 5.11. Under an external magnetic excitation, the stress applied to the piezoelectric layer by the Terfenol-D one is asymmetric, resulting in a bending motions rather than a longitudinal one for the L–T mode. In general, the ME field coefficients of the bending mode are smaller (~1.2 V/cmOe) than those of the L–T or L–L ones, simply because the bending mode contains only a single Terfenol-D layer. However, the bending mode does have better low-frequency response to magnetic fields than either the L–T or L–L, making it suitable for low-frequency magnetic sensor applications. Figure 5.18 shows the induced charge response of the bimorph ME laminate (given in Fig. 5.11c) to low-frequency magnetic field variations. The ME charge coefficient of this bimorph was ~800 pC/Oe at $f = 10^{-2}$ Hz.

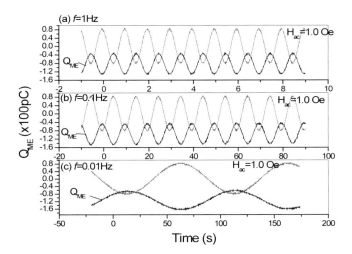

Figure 5.18 Extremely low-frequency responses of bimorph-type ME sensor. See also Color Insert.

5.2.5 C–C Terfenol-D/PZT and PZN–PT Laminates [35–37]

In many situations, magnetic fields are excited by electric currents. In this case, the excited magnetic fields are vortexes. A ME ring-type laminate operated in circumferential magnetization and circumferential polarization has been designated as the C–C mode, which is illustrated in Fig. 5.13. Experimental investigations have shown that Terfenol-D/PMN–PT/Terfenol-D three-layer laminated ring has a very high ME coefficient with maximum values of up to 5.5 V/cmOe at f = 1 kHz in response to a vortex magnetic field [37]. High ME coupling in the C–C mode is due to the magnetic-loop of the ring ME-type configuration, and which is suitable for capturing a vortex-type field. We will see later that this C–C ME laminate ring has potential for electric current sensor applications in non-resonance frequency ranges.

5.2.6 ME Laminates Based on Non-Terfenol-D Materials

Magnetostrictive $Tb_{1-x}Dy_xFe_{2-y}$ (Terfenol-D) has the highest magnetostriction amongst all known magnetostrictive materials. However, this rare earth alloy is quite costly, and also brittle. There are other magnetostrictive materials — such as Permendur, $Ni_{1-x}Co_xFe_2O_4$ (i.e., NFO), $Co_{1-x}Zn_xFe_2O_4$ (i.e., CFO), $LaSrMnO_3$ or $LaCaMnO_3$ etc. [16, 19, 26, 31–34, 59]. Recently, we also developed new ME laminate composites of magnetostrictive Fe-20at%Ga alloys with piezoelectric PZT ceramics and piezoelectric single crystals, which exhibited a large ME coupling [38, 39]. The maximum ME voltages observed for L–T Fe-Ga/PMN–PT laminates was ~1 V/cmOe at f = 1 kHz and 70 V/cmOe at resonance, see Fig. 5.19. Although low-frequency ME performance of Fe-Ga/PMN–PT laminates is not as good as that of Terfenol-D/PZT laminates, its resonance ME performance is better than that of Terfenol-D/PZT laminates, due to a higher Q-factor.

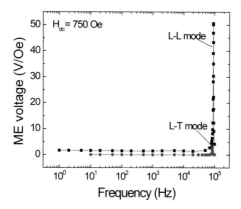

Figure 5.19 ME voltage coefficients of Fe-Ga/PMN–PT laminates as a function of magnetic field frequency [39]. See also Color Insert.

A most attractive alternative magnetostrictive material is a ribbon alloy, FeBSiC or Metglas. We will show our recent progress [40] on the use of this material in ME laminates later, in a section entitled Future Directions.

5.2.7 Three-Phase High-μ Ferrite/Terefenol-D/PZT Composites [41, 42]

In prior chapters, we have seen that two-phase magnetostrictive/ piezoelectric laminates or composites exhibit large ME coupling only near an optimum dc magnetic bias $H_{dc,opti}$, where the effective piezomagnetic coefficient (i.e., $d\lambda/dH$) of the magnetostrictive layers is maximum [22, 38]. Typical values of $H_{dc,opti}$ for L–T and L–L ME laminates are about 500 Oe, whereas those operated in T–T modes may have $H_{dc,opti}$ as high as 4000 Oe [17, 24]. These relatively high $H_{dc,opti}$ are due to the low magnetic permeability μ_r of the magnetostrictive materials that previously were used in two-phase ME composites. Typically, μ_r in magnetostrictive Terfenol-D or Fe-Ga is as low as 3 to 10. However, by incorporating a third phase ferromagnetic layers with a high permeability into ME Terfenol-D/PZT laminates, the effective permeability of ME composites can be dramatically increased, which in turn will result in a larger effective piezomagnetic or magnetostrictive coefficient in the magnetostrictive Terfenol-D and correspondingly a stronger ME coupling at lower $H_{dc,opti}$. In prior reports [41, 42], it has been found that $H_{dc,opti}$ of ME laminates can be dramatically decreased via co-lamination with a third phase high-permeability mu-metal or $MnZnFe_2O_4$ ferrite layers.

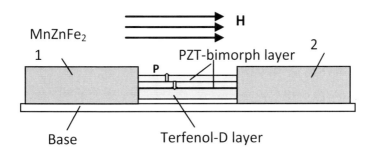

Figure 5.20 Configuration of three-phase $MnZnFe_2O_4$/Terfenol-D/PZT ME laminate [42]. See also Color Insert.

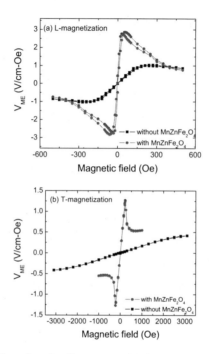

Figure 5.21 High-µ phase's effect on ME voltage coefficients in three-phase ME composites. See also Color Insert.

This, in turn, resulted in a significant enhancement of the apparent ME voltage coefficient by a factor up to 28× in low dc magnetic bias ranges. Figure 5.20 illustrates the configuration of three-phase $MnZnFe_2O_4$/Terfenol-D/PZT ME laminate and Fig. 5.21 shows the effect of third phase $MnZnFe_2O_4$ layers on ME voltage coefficient. A reduced $H_{dc,opti}$ is important for practical applications of ME materials.

5.3 ME LOW-FREQUENCY DEVICES

The working principle of magnetic sensing in ME laminates is simple: when a magnetic field is applied to the magnetostrictive layer, it strains, producing a proportional charge in the piezoelectric layer. Highly sensitive magnetic field sensors can be obtained using large magnetostrictive and high strain-sensitive piezoelectric ME composite materials in long- or ring-type configurations.

5.3.1 AC Magnetic Field Sensors

Figure 5.22 shows the voltages induced across the two ends of a PMN–PT layer in a push–pull ME Terfenol-D/PMN–PT/Terfenol-D three-layer laminate as a function of ac magnetic field (H_{ac}) at drive frequencies of (i) f = 1 kHz and (ii) f = 77.5 kHz under a given H_{dc}. In this figure, the induced ME voltage can be seen to have a linear response to H_{ac} over a wide range of fields from $10^{-11} < H_{ac} < 10^{-3}$ T [13, 22, 29]. When the laminates were operated under resonance drive, an enhancement in sensitivity to small magnetic field variations was observed. The sensitivity limit of ME laminates at ambient conditions was 1.2×10^{-12} T [2]. These results unambiguously demonstrate that ME laminates have an ultra-high sensitivity to small ac magnetic field variations.

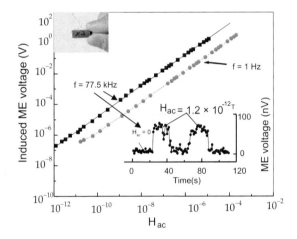

Figure 5.22 Limit magnetic field sensitivity of Terfenol-D/PMN–PT ME laminate [22]. See also Color Insert.

5.3.1.1 Extremely low-frequency magnetic field sensors [33, 34]

Apart from a bimorph [29], a multilayer configuration [33,34] of ME laminates has been reported that enables ultra-low-frequency detection of magnetic field variations This multilayer ME laminate is illustrated earli er in Fig. 5.10h. This configuration can greatly improve the low-frequency capability because of its high ME charge coupling and large capacitance. Figure 5.23 shows the magnetic field induced voltages for multilayer Terfenol-D/PMN–PT ME laminates at frequencies of 100 Hz, 1 Hz, and 0.01 Hz. At an extremely low frequency of f = 10 mHz, multilayer ME laminates can still detect a small magnetic field variation as low as 10^{-7} T.

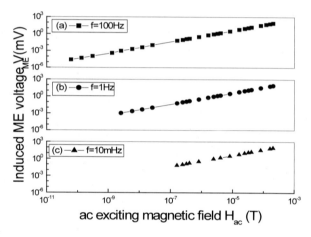

Figure 5.23 Low-frequency response of a multi-L–T Terfenol-D/PMN–PT laminate [33].

5.3.1.2 DC magnetic field sensors [21, 43]

It has also been reported that small dc magnetic field variations can be detected using ME laminates based on magnetic bias effect [21] or inverse ME effect [43, 44]. In fact, small long-type ME laminates of Terfenol-D and PZT are quite sensitive to small H_{dc} variations, when driven under a constant H_{ac}. The sensitivity limit is about 10^{-7} T using a constant amplitude low frequencies drive, which can be enhanced to 10^{-8} T under resonant drive [21]. Figure 5.24 shows the sensitivity limit of an L–T ME laminate to small dc magnetic field variations, while under resonant drive. It can be seen that dc magnetic field changes as small as 10^{-8} T were readily detected.

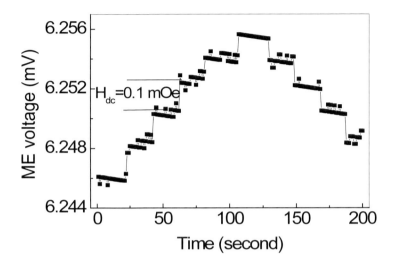

Figure 5.24 Sensitivity limit of an L–T ME laminate to small dc magnetic field variations under resonant drive [21].

5.3.2 ME Current Sensors

ME composites have also been shown to be good candidates for electric current sensors. A straight wire containing ac or dc current I will excite an ac or dc vortex magnetic field H_{vor} around this wire: $H_{vor} = I/\pi r$, where r is radius of the vortex magnetic field. Accordingly, ring-type ME laminates [35–37] are ideal configurations for vortex magnetic field detection, or current I detection. Previously, a ME ring-type laminate has been made of a circumferentially magnetized Terfenol-D and a circumferentially poled piezoelectric PZT (or PMN–PT), which was shown to have a high sensitivity of up to a_{ME} = 5.5 V/cmOe at f = 1 kHz to a vortex magnetic field. Figure 5.25a illustrates a ring-type ME laminate used as an electric current sensor, and Fig. 5.25b shows the ME voltage response to a square-wave current passing through the wire. Detection using a toroidal-type variable reluctance coil (100 turns) exhibited a much smaller induced voltage (by a factor of 0.01 times) than that of our ME ring sensor.

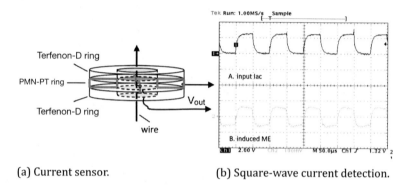

(a) Current sensor.   (b) Square-wave current detection.

Figure 5.25 ME current sensor [37]. See also Color Insert.

5.3.3 ME Transformers and Gyrators

ME transformers or gyrators have important applications as voltage gain devices, current sensors, and other power conversion devices.

An extremely high voltage gain effect under resonance drive has been reported in long-type ME laminates consisting of Terfenol-D and PZT layers [14, 45, 46]. A solenoid with n turns around the laminate that carries a current of I_{in} was used to excite a H_{ac}. The input ac voltage applied to the coils was V_{in}. When the frequency of H_{ac} was equal to the resonance frequency of the laminate, the ME voltage coefficient was strongly increased, and correspondingly the output ME voltage (V_{out}) induced in the piezoelectric layer was much higher than V_{in}. Thus, under resonant drive, ME laminates exhibit a strong voltage gain, offering potential for high-voltage miniature transformer applications.

Figures 5.26 and 5.27 show the measured voltage gain V_{out}/V_{in} as a function of the drive frequency f for the ME transformer and gyration consisting of Terfenol-D layers 40 mm in length and a piezoelectric layer 80 mm in length. A maximum voltage gain of ~260 was found at a resonance frequency of 21.3 kHz. In addition, at the resonance state, the maximum voltage gain of the ME transformer was strongly dependent on an applied H_{dc}, which was due to the fact that Terfenol-D has a large effective piezomagnetic coefficient only under a suitable H_{dc}.

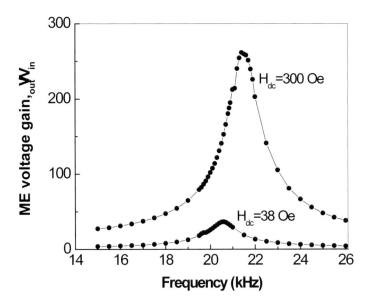

Figure 5.26 Voltage gain of ME transformer as a function of the drive frequency [3].

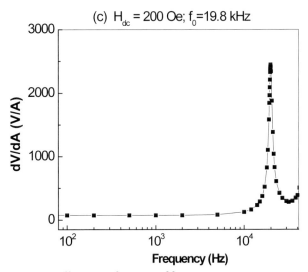

(a) *I–V* gyration coefficient as function of frequency.

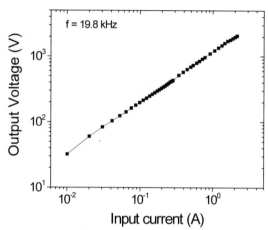

(b) *I–V* gyration at resonance frequency of 19.8 kHz.

Figure 5.27 *I–V* gyration [47].

5.4 FUTURE DIRECTIONS

5.4.1 Terfenol-D-Based Composites

Terenol-D-based ME composites have shown strong ME coupling over wide frequency range. Further optimizations for Terfenol-D-based ME laminate composites should focus on:

i. Multilayer long or ring-type configurations for structure optimization;
ii. Optimized magnetic–piezoelectric phase ratio;
iii. Incorporation of even higher-μ third phases;
iv. Engineering controlled connectivity of the magnetic–piezoelectric phases.

It is also important to further enhance the effective piezomagnetic coefficient of the magnetostrictive phases and to lower its required magnetic bias, and develop piezoelectric materials with high piezoelectric voltage constant $g_{33,p}$. Recently, significant progress has been made in this goal, as will be described below.

5.4.2 Metglas/PZT Fiber (2–1) Composites

As discussed above, the maximum ME voltage coefficient (subresonant) has been previously been reported to be 5 V/cmOe [37], and generally it has been found to be notably smaller than this (≤1 V/cmOe); although, resonance enhancement to values of 70–90 V/cmOe have been reported. Recently, Metglas ribbons have been used as the magnetostrictive layers [40]. Metglas has an extremely high magnetic permeability ($\mu_r \sim$ 30,000), but low magnetostriction. However, because of the ultra-high μ_r, the effective piezomagnetic coefficient is high under small dc magnetic bias, making it an ideal candidate for incorporation into ME composites. In addition, prior research [4] also has shown a piezoelectric layer cut into fibers has enhanced higher piezoelectric voltage coefficient $g_{33,p}$, relative to its monolithic forms. Accordingly, a two-dimension Metglas layer connected with a one-dimension piezoelectric PZT fiber layer (i.e., a (2–1) connectivity Metglas/PZT fiber laminate) has recently been shown to have a much enhanced ME coupling. Figure 5.28 illustrates the ME laminate design for this Metglas/PZT-fiber laminate.

Figure 5.29 shows the reported ME field coefficient(dE/dH) as a function of dc magnetic bias for longitudinal (L), width ($T1$) and thickness ($T2$) magnetizations. From this figure, it can be seen that the maximum dE/dH for the L–L mode is 22 V/cmOe under a H_{dc} of only 5 Oe. This is almost an order higher than that of corresponding L–L mode of Terfenol-D-based ME laminates. Furthermore, under resonance drive, this value was can be enhanced to ~500 V/cmOe. The data in Fig. 5.29 also show strong anisotropy: the width and thickness direction of the laminate had a dE/dH 100× less than that along the longitudinal direction. Clearly, Metglas/PZT fibers in (2–1) connectivity are an important direction for new ME laminate developments.

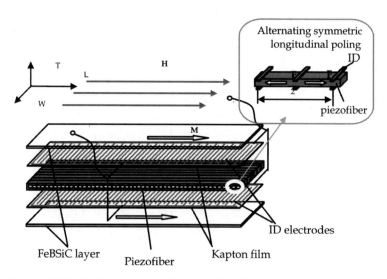

Figure 5.28 (2–1) connectivity Metglas/PZT fiber laminate [40]. See also Color Insert.

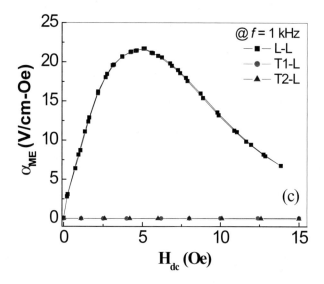

Figure 5.29 ME field coefficient as a function of dc magnetic bias for longitudinal (L), width ($T1$), and thickness ($T2$) magnetization [40]. See also Color Insert.

5.5 CONCLUSIONS

In this chapter, the basic working modes of the ME laminates, i.e., longitudinally magnetized and transversely polarized (or L–T), longitudinally magnetized and longitudinally polarized (or L–L), and transversely magnetized and transversely polarized (or T–T), modes were defined, and correspondingly, the magneto–elasto–electric equivalent circuits for these modes were developed. The proposed equivalent circuit method is quite useful for both static and dynamic analysis of ME laminates, especially for electromechanical resonance analysis. Our equivalent circuit method predicts that (i) ME coupling in ME laminates is strongly related to their working modes: the value of dV/dH in L–T or L–L mode is significantly higher relative to the T–T, (ii) there is an optimum thickness ratio of piezoelectric and magnetostrictive layers that maximizes dV/dH, (iii) resonance ME voltage coefficient (dV/dH) (ω_s) is $\sim Q_m$ times higher under resonance drive. These predictions have been confirmed by experimental results.

In low-frequency magnetic field applications, ME laminates made of magnetostrictive Terfenol-D alloy and piezoelectric PZT ceramics or PMN–PT single crystals have shown promise applications, such as high sensitivity magnetic sensors, magnetic compasses, ME transformers, current sensors, or gyrators. Recently, ME laminates made of high-permeability FeBSiC foils and PZT or PZN–PT fibers have even shown much higher ME coupling coefficients: ~22 V/cmOe at f = 1 kHz or ~500 V/cmOe at electromechanical resonance, which is almost one order of magnitude higher than that of Terfenol-D/PZT laminates. New applications of ME laminates at low frequency may further include magnetic, acoustic, or vibration energy harvesting due to their strong magneto–elasto–electric coupling ability.

Acknowledgments

This work was supported by the National Basic Research Program of China, the Office of Naval Research, U.S. Department of Energy, and DARPA.

References

1. H. Schmid, "On a magnetoelectric classification of materials," Proc. Symposium on Magnetoelectric Interaction in Crystals, USA, 1973, Eds. A. Freeman and H. Schmid H., Gordon and Breach Sci. Publ., New York, **121** (1975).

2. R. Fuchs, "Wave propagation in a magnetoelectric medium," Phyl. Mag., **11**, 647 (1965).

3. R. Rottenbacher, H.J. Oel, G. Tomandel, "Ferroelectrics ferromagnetics," Ceramics Int., **106**, 106 (1981).

4. G. Harshe, J.O. Dougherty, R.E. Newnham, "Theoretical modelling of multilayer magnetoelectric composites," Int. J. Appl. Electromagn. Mater., **4**, 145 (1993).

5. G. Srinivasan, E.T. Rasmussen, J. Gallegos, R. Srinivasan, Yu.I. Bokhan, V.M. Laletin, "Novel magnetoelectric bilayer and multilayer structures of magnetostrictive and piezoelectric oxides," Phys. Rev., **B64**, 214 (2001).

6. O.V. Rybkov, V.M. Petrov, S.V. Averkin, G. Srinivasan, "Magnetoacoustic resonance in ferrite-piezoelectric bilayer structures subject to exchange interaction," Bulletin of NovSU, **39**, 110 (2006) (in Russian).

7. B.D.H. Tellegen, "The gyrator, a new electric network element," Philips Res. Rep., **3**, 81 (1948).

8. J.G. Wan, J.-M. Liu, H.L.W. Chand, C.L. Choy, G.H. Wang, C.-W. Nan, "Giant magnetoelectric effect of a hybrid of magnetostrictive and piezoelectric composites," J. Appl. Phys., **93**, 9916 (2003).

9. S. Dong, J.F. Li, D. Viehland, "Giant magneto-electric effect in laminate composites," IEEE Trans. Ultrason. Ferroelectr. Freq. Control, **50**, 1236 (2003).

10. S. Dong, J.F. Li, D. Viehland, "Characterization of magnetoelectric laminate composites operated in longitudinal-transverse and transverse-transverse modes," J. Appl. Phys., **95**, 2625 (2004).

11. K. Mori, M. Wuttig, "Magnetoelectric coupling in terfenol-d/polyvinylidenedifluoride composites," Appl. Phys. Lett., **81**, 100 (2002).

12. C.-W. Nan, "Magnetoelectric effects in composites of piezoelectric and piezomagnetic phases," Phys. Rev., **B50**, 6082 (1994).

13. S. Dong, J.F. Li, D. Viehland, "Ultrahigh magnetic field sensitivity in laminate of Terfenol-D and Pb(Mg$_1$/3Nb$_2$/3)O$_3$-PbTiO$_3$ crystals," Appl. Phys. Lett., **83**, 2265 (2003).

14. S. Dong, J.F. Li, D. Viehland, "Magneto-electric coupling, efficiency and voltage gain effect in piezoelectric-piezomagnetic laminate composite: theory and analysis," J. Mater. Sci., **41**, 97 (2006).

15. M. Avellaneda, G. Harshe, "Magnetoelectric effect in piezoelectric/magnetostrictive multiplayer (2-2) composites," J. Intell. Mater. Syst. Struct., **5**, 501 (1994).

16. J. Wu, J. Huang, "Closed-form solutions for the magnetoelectric coupling coefficients in fibrous composites with piezoelectric and piezomagnetic phases," Int. J. Solids Struct., **37**, 2981 (2000).

17. S. Dong, J.F. Li, D. Viehland, "Longitudinal and transverse magneto-electric voltage coefficients of magnetostrictive/piezoelectric laminate composite: theory," Ultrason. Ferroelectr. Freq. Control, **50**, 1253 (2003).

18. S. Dong, J. Cheng, J.F. Li, D. Viehland, "Enhanced magneto-electric effects in laminates of Terfenol-D/PZT under resonance drive," Appl. Phys. Lett., **83**, 4812 (2003).

19. M. Avellaneda, G. Harshe, "Magnetoelectric effect in piezoelectric/magnetostrictive multiplayer (2-2) composites," J. Intell. Mater. Syst. Struct., **5**, 501 (1994).

20. J. Ryu, A.V. Carazo, K. Uchino, H.E. Kim, "Magnetoelectric properties in piezoelectric and magnetostrictive composites," Jpn. J. Appl. Phys., **40**, 4948 (2001).

21. S. Dong, J. Zhai, J.F. Li, D. Viehland, "Small dc magnetic field response of magnetoelectric laminate composites," Appl. Phys. Lett., **88**, 082907 (2006).

22. S. Dong, J. Zhai, F. Bai, J.F. Li, D. Viehland, "Push-pull mode magnetostrictive/piezoelectric laminate composite with an enhanced magnetoelectric voltage coefficient," Appl. Phys. Lett., **87**, 062502 (2005).

23. S. Dong, J.F. Li, D. Viehland, "A longitudinal longitudinal mode TERFENOL-D/Pb(Mg$_1$/3Nb$_2$/3)O$_3$–PbTiO$_3$ laminate composite," Appl. Phys. Lett., **85**, 5035 (2004).

24. S. Dong, J.F. Li, D. Viehland, "Longitudinal and transverse magneto-electric effect: II. experiments," Ultrason. Ferroelectr. Freq. Control, **51**, 794 (2004).

25. D.A. Filippov, M.I. Bichurin, C.-W. Nan, J.M. Liu, "Magnetoelectric effect in hybrid magnetostrictive-piezoelectric composites in the electromechanical resonance region," J. Appl. Phys., **97**, 113910 (2005).

26. S. Stein, M. Wuttig, D. Viehland, E. Quandt, "Magnetoselectric effect in sputtered composites," J. Appl. Phys., **97**, 10Q301 (2005).

27. J. Zhai, S. Dong, Z. Xing, J.F. Li, D. Viehland, "Giant magneto-electric effect in PVDF/Metglas laminates," Appl. Phys. Lett., **89**, 083507 (2006).

28. Z. Xing, S. Dong, J. Zhai, J.F. Li, D. Viehland, "A resonant bending-mode of magneto-electric laminate composites of Terfenol-D/Steel/Pb(Zr,Ti)O$_3$," Appl. Phys. Lett., **89**, 112911 (2006).

29. J. Zhai, Z. Xing, S. Dong, J.F. Li, D. Viehland "Detection of pico-Tesla magnetic fields using magneto-electric sensors at room temperature," Appl. Phys. Lett., **88**, 062510 (2006).

30. J.G. Wan, Z.Y. Li, Y. Wang, M. Zeng, G.H. Wang, J.-M. Liu, "Strong flexural resonant magnetoelectric effcet in Terfenol-D/epoxy-Pb(Zr,Ti)O$_3$ bilayer," Appl. Phys. Lett., **86**, 202504 (2005).

31. G. Srinivasan, E.T. Rasmussen, B.J. Levin, R. Hayes, "Magnetoelectric effect in bilayers and multilayers of magnetostrictive and piezoelectric perovskite oxides," Phys. Rev., **B65**, 134402 (2002).

32. M.I. Bichurin, D.A. Filippov, V.M. Petrov, V.M. Laletsin, N. Paddubnaya, G. Srinivasan, "Resonance magnetoelectric effects in layered magnetoelectric-piezoelectric composites," Phys. Rev., **B68**, 132408 (2003).

33. S. Dong, J. Zhai, Z. Xing, J.F. Li, D. Viehland, "Extremely low frequency response of magnetoelectric multilayer composites," Appl. Phys. Lett., **86**, 102901 (2005).

34. K.E. Kamentsev, Y.K. Fetisov, G. Srinivasan, "Low-frequency nonlinear magnetoelectric effects in a ferrite-piezoelectric multilayer," Appl. Phys. Lett., **89**, 142510 (2006).

35. S. Dong, J.F. Li, D. Viehland, "Circumferentially magnetized and circumferentially polarized magnetostrictive/piezoelectric laminated rings," J. Appl. Phys., **96** 3382 (2004).

36. S. Dong, J.F. Li, D. Viehland, "Vortex magnetic field sensor based on ring-type magnetoelectric laminate," Appl. Phys. Lett., **85**, 2307 (2004).

37. S. Dong, J.G. Bai, J. Zhai, J.F. Li, G.-Q. Lu, D. Viehland, S. Zhang, T.R. Shrout, "Circumferential-mode, quasi-ring-type, magnetoelectric laminate composite — a highly sensitive electric current and/or vortex magnetic field sensor," Appl. Phys. Lett., **86**, 182506 (2005).

38. S. Dong, J. Zhai, F. Bai, J.F. Li, D. Viehland, T.A. Lograsso, "Magnetostrictive and magnetoelectric behavior of Fe-20 at.% Ga/Pb(Zr,Ti) O3 laminates," J. Appl. Phys., **97**, 103902 (2005).

39. S. Dong, J. Zhai, Naigang Wang, Feiming Bai, J.F. Li, D. Viehland, T.A. Lograsso, "Fe-Ga/Pb(Mg$_1$/3Nb$_2$/3)O$_3$-PbTiO$_3$ magnetoelectric laminate composites," Appl. Phys. Lett., **87**, 222504 (2005).

40. S. Dong, J. Zhai, J.F. Li, D. Viehland, "Near-perfect magnetoelectricity in high-permeability magnetostrictive/piezoelectric-fiber laminates," Appl. Phys. Lett., **89**, 252904 (2006).

41. S. Dong, J. Zhai, J.F. Li, D. Viehland, "Magneto-electric effect in Terfenol-D/PZT/mu-metal composite," Appl. Phys. Lett., **89**, 122903 (2006).

42. S. Dong, J. Zhai, J.F. Li, D. Viehland, "Enhanced magnetoelectric effect in three-phase MnZnFe$_2$O$_4$/Tb$_{1-x}$Dy$_x$Fe$_{2-y}$/Pb(Zr,Ti)O$_3$ composites," J. Appl. Phys., **100**, 124108 (2006).

43. J.G. Wan, J.-M. Liu, G.H. Wang, C.-W. Nan, "Electric-field-induced magnetization in Pb(Zr,Ti)O$_3$/Terfenol-D composite structure," Appl. Phys. Lett., **88**, 182505 (2006).

44. Z. Huang, "Theoretical moeling on the magnetization by electric field through product property," J. Appl. Phys., **100**, 114104 (2006).

45. S. Dong, J.F. Li, D. Viehland, "A strong magnetoelectric voltage gain effect in magnetostrictive-piezoelectric composite," Appl. Phys. Lett., **85**, 3534 (2004).

46. S. Dong, J.F. Li, D. Viehland, "Voltage gain effect in a ring-type magnetoelectric laminate," Appl. Phys. Lett., **84**, 4188 (2004).

47. S. Dong, J.Y. Zhai, J.F. Li, D. Viehland, M.I. Bichurin, "Magnetoelectric gyration effect in Tb$_{1-x}$Dy$_x$Fe$_{2-y}$/Pb(Zr,Ti)O$_3$ laminated composites at the electromechanical resonance," Appl. Phys. Lett., **89**, 243512 (2006).

48. J. Zhai, J.F. Li, S. Dong, D. Viehland, M.I. Bichurin, "A quasi (unidirectional) tellegen gyrator," J. Appl. Phys., **100**, 124509 (2006).

49. Cid-Pastor, L. Martinez-Salamero, C. Alonso, B. Estibals, J. Alzieu, G. Schweitz, D. Shmilovitz, "Analysis and design of power gyrators in sliding-mode operation," IEE Proc. Elec. Power Appl., **152**, 821 (2005).

50. A. Zeki, A. Toker, "DXCCII-based tunable gyrator," Int. J. Electron. Commun. (AEU) **59**, 59 (2005).

51. S. Franco, *Design with operational amplifiers and analog integrated circuits*, 3rd ed., Tata McGraw Hill, New Delhi, 60 (2002).

52. B. Guthrie, J. Hughes, T. Sayers, A. Spencer, "A CMOS gyrator low-IF filter for a dual-mode bluetooth/zigbee transceiver," IEEE J. Solid-State Circuits, **40**, 1872 (2005).

53. A. Morse, L. Huelsman, "A gyrator realization using operational amplifiers," IEEE Trans. Circuit Theory, CT-11, 277 (1964).

54. A. Antoniou, "Realisation of gyrators using operational amplifiers, and their use in TfC-active-network synthesis," Proc. Inst. Elec. Eng., **116**, 1838 (1969).

55. M. Ehsani, I. Husain, M.O. Bilgic, "Power converters as natural gyrators," IEEE Trans. Circ. Syst. Fund. Theor. Appl., **40**, 946 (1993).

56. C.L. Hogan, "The ferromagnetic faraday effect at microwave frequencies and its applications," Rev. Mod. Phys., **25**, 253 (1953).

57. J. Mazur, M. Solecka, M. Mazur, R. Poltorak, E. Sedek, "Design and measurement of gyrator and isolator using ferrite coupled microstrip lines," IEE Proc. Microw. Antennas Propag., **152**, 43 (2005).

58. E. Asher, "The interaction between magnetization and polarization: phenomenological symmetry consideration," J. Phys. Soc. Jpn., **28**, 7 (1969).

59. V.M. Laletsin, N.N. Padubnaya, G. Srinivasan, C.P. DeVreugd, "Frequency dependence of magnetoelectric interactions in layered structures of ferromagnetic alloys and piezoelectric oxides," Appl. Phys., **A78**, 33 (2004).

Chapter 6

Ferrite–Piezoelectric Composites at Ferromagnetic Resonance Range and Magnetoelectric Microwave Devices

G. Srinivasan[a] and M.I. Bichurin[b]

[a]*Department of Physics, Oakland University, Rochester,*
Michigan 48309-4401, USA
[b]*Institute of Electronic and Information Systems,*
Novgorod State University,
173003 Veliky Novgorod, Russia

In this chapter, we present a detailed treatment for electric field-induced resonance field shift for ferromagnetic resonance in multilayers and a theoretical analysis of high-frequency magnetoelectric (ME) effects for a ferrite–piezoelectric bilayer. Resonant dependence of the ME voltage coefficient is found out at overlap the lines of electromechanical and ferromagnetic resonances. In addition, similiar ME phenomena are considered that do not require bonding between the layers and take place simply due to the proximity of two material having different dielectric and magnetic properties. Both *H*- and *E*-dependence of

Magnetoelectricity in Composites
Edited by Mirza I. Bichurin and Dwight Viehland
Copyright © 2012 Pan Stanford Publishing Pte. Ltd.
www.panstanford.com

hybrid excitations, due to variations in permeability and permittivity, were investigated.

Ferrite–piezoelectric composites represent a promising new approach to building a new class of fast electric field tunable low power devices based on ME interactions. The ME composites are ready for technological applications. Because of the shift in the resonance frequency in a static magnetic or electric bias field the composite materials hold promise in electrically tunable microwave applications such as filters, oscillators, and phase shifters.

The majority of prior work on magnetoelectric (ME) composites, both theoretical and experimental studies, has been devoted to the low-frequency range (10 Hz–10 kHz) [1–6]. However, ME composites also offer important applications in the microwave range [8]. In this frequency range, the ME effect reveals itself as a change in the magnetic permeability under an external electric field. Investigations of a ferromagnetic resonance (FMR) line shift by an applied electric field are easily performed for layered ferrite–piezoelectric structures [10–15]. In addition, layered composites are of interest for applications as electrically tunable microwave phase shifters, devices based on FMR, magnetic-controlled electrooptical and/or piezoelectric devices, and electrically readable magnetic (ME) memories [28–30].

In this chapter, we discuss theoretical models of multilayer ME composites in the microwave range [14–16]. The model is a phenomenological theory, which is based on the magnetic susceptibility as function of strain [12], whose this strain-dependence is defined both by piezoelectric and the magnetoelastic constants of piezoelectric and magnetostrictive phases, respectively.

6.1 BILAYER STRUCTURE

Let us again consider the simple model of a bilayer structure, consisting of a ferrite spinel with cubic (m3m) symmetry and poled lead zirconate titanate (PZT) with a symmetry of ∞m [16]. The influence of an electric field on the piezoelectric PZT phase can be described as follows:

$$^{P}T_{ij} = {}^{P}c_{ijkl}{}^{P}S_{kl} - e_{kij}E_{k},$$ (6.1)

where E_{k} is a component of the electric field vector; and $^{P}T_{ij}$, $^{P}S_{kl}$, e_{kij}, and $^{P}c_{ijk}$ are the components of the stress, strain, piezoelectric, and elastic stiffness tensors of the piezoelectric phase, respectively. The tensor coefficients of the elastic stiffness has the form

$$^{P}c = \begin{pmatrix} ^{P}A_{11} & ^{P}c_{12} & ^{P}c_{13} & 0 & 0 & 0 \\ ^{P}c_{12} & ^{P}c_{11} & ^{P}c_{13} & 0 & 0 & 0 \\ ^{P}c_{13} & ^{P}c_{13} & ^{P}c_{11} & 0 & 0 & 0 \\ 0 & 0 & 0 & ^{P}c_{44} & 0 & 0 \\ 0 & 0 & 0 & 0 & ^{P}c_{44} & 0 \\ 0 & 0 & 0 & 0 & 0 & \frac{1}{2}\left(^{P}c_{11} - {}^{P}c_{12}\right) \end{pmatrix}$$ (6.2)

and that of the piezoelectric coefficients is by 0

$$e = \begin{pmatrix} 0 & 0 & 0 & 0 & e_{15} & 0 \\ 0 & 0 & 0 & e_{15} & 0 & 0 \\ e_{31} & e_{31} & e_{33} & 0 & 0 & 0 \end{pmatrix}$$ (6.3)

Substituting the Eqs. 6.2 and 6.3 into Eq. 6.1, and by applying an electric field along the axis of polarization (i.e., $E_{3} = E$, $E_{1} = E_{2} = 0$), we obtain the following stresses in the piezoelectric phase:

$$\begin{cases} {}^P T_{33} = -e_{33}E + {}^P c_{13}({}^P S_{11} + {}^P S_{22}) + {}^P c_{33}\,{}^P S_{33}, \\ {}^P T_{11} = -e_{31}E + {}^P c_{11}\,{}^P S_{11} + {}^P c_{12}\,{}^P S_{22} + {}^P c_{13}\,{}^P S_{33} = 0, \\ {}^P T_{22} = -e_{31}E + {}^P c_{12}\,{}^P S_{11} + {}^P c_{11}\,{}^P S_{22} + {}^P c_{13}\,{}^P S_{33} = 0. \end{cases} \quad (6.4)$$

Now, assume that the axis of polarization in the piezoelectric phase coincides with the crystallographic [111] axis of the magnetostrictive phase. Then, the elasticity tensor of the magnetostrictive phase becomes

$$
{}^m c = \begin{pmatrix}
\frac{{}^m c_{11}+{}^m c_{12}}{2}+{}^m c_{44} & \frac{{}^m c_{11}+5{}^m c_{12}-2{}^m c_{44}}{6} & \frac{{}^m c_{11}+2{}^m c_{12}-2{}^m c_{44}}{3} & \frac{{}^m c_{11}-{}^m c_{12}-2{}^m c_{44}}{3\sqrt{2}} & 0 & 0 \\
\frac{{}^m c_{11}+5{}^m c_{12}-2{}^m c_{44}}{6} & \frac{{}^m c_{11}+{}^m c_{12}}{2}+{}^m c_{44} & \frac{{}^m c_{11}+2{}^m c_{12}-2{}^m c_{44}}{3} & \frac{{}^m c_{11}-{}^m c_{12}-2{}^m c_{44}}{3\sqrt{2}} & 0 & 0 \\
\frac{{}^m c_{11}+2{}^m c_{12}-2{}^m c_{44}}{3} & \frac{{}^m c_{11}+2{}^m c_{12}-2{}^m c_{44}}{3} & \frac{{}^m c_{11}+2{}^m c_{12}+4{}^m c_{44}}{3} & 0 & 0 & 0 \\
\frac{{}^m c_{11}+2{}^m c_{12}-2{}^m c_{44}}{3} & -\frac{{}^m c_{11}-{}^m c_{12}-2{}^m c_{44}}{3\sqrt{2}} & 0 & \frac{{}^m c_{11}-{}^m c_{12}+{}^m c_{44}}{3} & 0 & 0 \\
0 & 0 & 0 & 0 & \frac{{}^m c_{11}-{}^m c_{12}+{}^m c_{44}}{3} & \frac{{}^m c_{11}-{}^m c_{12}-2{}^m c_{44}}{3} \\
0 & 0 & 0 & 0 & \frac{{}^m c_{11}-{}^m c_{12}-2{}^m c_{44}}{3\sqrt{2}} & \frac{{}^m c_{11}-{}^m c_{12}+4{}^m c_{44}}{6}
\end{pmatrix}
$$

$$(6.5)$$

The stresses in the magnetostrictive phase considering are then given by

$$
\begin{cases}
{}^m T_1 = \left(\frac{{}^m c_{11}+{}^m c_{12}}{2}+{}^m c_{44}\right){}^m S_1 + \frac{{}^m c_{11}+5{}^m c_{12}-2{}^m c_{44}}{6}\,{}^m S_2 + \frac{{}^m c_{11}+2{}^m c_{12}-2{}^m c_{44}}{3}\,{}^m S_3 + \frac{{}^m c_{11}-{}^m c_{12}-2{}^m c_{44}}{3\sqrt{2}}\,{}^m S_4, \\[2ex]
{}^m T_2 = \frac{{}^m c_{11}+5{}^m c_{12}-2{}^m c_{44}}{6}\,{}^m S_1 + \left(\frac{{}^m c_{11}+{}^m c_{12}}{2}+{}^m c_{44}\right){}^m S_2 + \frac{{}^m c_{11}+2{}^m c_{12}+4{}^m c_{44}}{3}\,{}^m S_3, \\[2ex]
{}^m T_3 = \frac{{}^m c_{11}+2{}^m c_{12}-2{}^m c_{44}}{3}\,{}^m S_1 + \frac{{}^m c_{11}+2{}^m c_{12}-2{}^m c_{44}}{3}\,{}^m S_2 + \frac{{}^m c_{11}+2{}^m c_{12}-2{}^m c_{44}}{3}\,{}^m S_3 - \frac{{}^m c_{11}-{}^m c_{12}-2{}^m c_{44}}{3\sqrt{2}}\,{}^m S_4, \\[2ex]
{}^m T_4 = \frac{{}^m c_{11}+2{}^m c_{12}-2{}^m c_{44}}{3}\,{}^m S_1 - \frac{{}^m c_{11}-{}^m c_{12}-2{}^m c_{44}}{3\sqrt{2}}\,{}^m S_2 + \frac{{}^m c_{11}-{}^m c_{12}+{}^m c_{44}}{3}\,{}^m S_4.
\end{cases}
$$

$$(6.6)$$

To calculate the ME effect of bilayer structures in the FMR range, we should use the following procedure: (i) ${}^m S_3$ is defined as a function of the stress ${}^m T_3$; (ii) ${}^P T_3$ is defined as a function of the strain ${}^P S_3$; and (iii) use the known expressions for the dependence of the resonant magnetic field on stress to determine the FMR line shift. Thus, the task is reduced to the solutions of electrostatic equations under specific boundary conditions.

Let us then consider that the bilayer structure is mechanically clamped along the 3 axis the ferrite and piezoelectric subsystems. In this case, the boundary conditions, without taking into account forces of friction, have the form

$$^{p}T_3 = {}^{m}T_3,$$

$$^{p}S_3 = -\frac{{}^{m}v}{{}^{p}v} \cdot {}^{m}S_3,$$

(6.7)

where ^{m}v and ^{p}v are the volume fractions of the magnetostrictive and piezoelectric phases, respectively. From Eqs. 6.4, 6.6, and 6.7 we can find that the stress along the 3 axis is given by

$$^{m}T_3 = \frac{E_3 \left(\dfrac{2\,{}^{p}c_{13}e_{31}}{{}^{p}c_{11} + {}^{p}c_{12}} - e_{33} \right)}{1 + \dfrac{{}^{m}v}{{}^{p}v}\left(\dfrac{1}{3\,{}^{m}c_{44}} + \dfrac{1}{3 + \left({}^{m}c_{11} + 2\,{}^{m}c_{12}\right)} \right)\left({}^{p}c_{33} - 2\dfrac{{}^{p}c_{13}^{\;2}}{{}^{p}c_{11} + {}^{p}c_{12}} \right)}.$$

(6.8)

It is known that an applied electric voltage results in a change of the resonant magnetic field [17]. We limit ourselves to the case when both the magnetic and electric fields are oriented along the polarization axis of the piezoelectric phases, which also corresponds to the [111] axis of the magnetostrictive one. In this case, the shift of the resonant magnetic field is given by

$$\delta H_E = \frac{3\lambda_{111}\,{}^{m}T_3}{M_0} = AE_3,$$

(6.9)

where M_0 is the saturation magnetization of the magnetostrictive layer, and the ME constant is defined by

$$A = \frac{3\lambda_{111}}{M_0} = \frac{\left(\dfrac{2\,{}^p c_{13} e_{31}}{{}^p c_{11} + {}^p c_{12}} - e_{33}\right)}{\left[1 + \dfrac{{}^m v}{{}^p v}\left(\dfrac{1}{3\,{}^m c_{44}} + \dfrac{1}{3 + \left({}^m c_{11} + 2\,{}^m c_{12}\right)}\right)\right]\left({}^p c_{33} - 2\dfrac{{}^p c_{13}{}^2}{{}^p c_{11} + {}^p c_{12}}\right)}.$$

$$(6.10)$$

From Eqs. 6.8 and 6.9, it is possible to estimate the shift of the resonant field for a bilayer magnetostrictive–piezoelectric composite. The theoretical values for the ME constant $A = \delta H_E/E_3$ were determined for composites of both nickel ferrite (NFO)–PZT and yttrium iron garnet (YIG)–PZT. For these calculations, the following materials parameters of two component phases were used: $YIG\,{}^p c_{11} = 12.6 \times 10^{10}\,\text{N/m}^2$, ${}^p c_{12} = 7.95 \times 10^{10}\,\text{N/m}^2$, ${}^p c_{13} = 8.4 \times 10^{10}\,\text{N/m}^2$, ${}^p c_{33} = 11.7 \times 10^{10}\,\text{N/m}^2$, $e_{31} = -6.5\,\text{Sim/m}^2$, and $e_{33} = 23.3\,\text{Sim/m}^2$: $NiFe_2O_4\,{}^m c_{11} = 22 \times 10^{10}\,\text{N/m}^2$, ${}^m c_{12} = 10.9 \times 10^{10}\,\text{N/m}^2$, ${}^m c_{44} = 8.12 \times 10^{10}\,\text{N/m}^2$, $\lambda_{111} = -21.6 \times 10^{10}$, and $4\pi M_0 = 3200$ G.

In Fig. 6.1, the dependence of the ME constant A on the relative volume fraction of the component phases in the bilayer structure is shown. In a composite of NFO and PZT, the ME effect has been reported to be stronger than that of YIG–PZT. This is because NFO has a much higher magnetostriction than YIG. Another notable feature in Fig. 6.1 is the fact that theoretically the field shift should increase with increasing volume fraction of the piezoelectric phase. However, please bear in mind that the magnetic resonance line will become to weak if the volume fraction of the magnetostrictive phase is too low. This is uniquely different than results at lower frequencies, where the largest ME effect is found for an approximate volume fraction ratio of 50:50 [7, 18]. Assuming that ${}^m v/{}^p v = 1$, we can estimate that the resonance line shift under applied electric field is given by the relation $\delta H = AE$, where $A \approx 2\,\text{Oe cm/kV}$ for $NiFe_2O_4$–PZT and $A \approx 0.45$ Oe-cm/kV for YIG–PZT.

The values for the calculated resonance line shift are in good agreement with prior experimental data [10]: theoretical line shift is 0.45 Oe-cm/kV, which is comparable to the measured one of 0.2–0.56 Oe-cm/kV depending on relative volume fraction of magnetostrictive and piezoelectric phases. Thus, we can see that it is necessary to use a piezoelectric component with large piezoelectric coefficients, and a magnetostrictive one with small saturation magnetization and high magnetostriction.

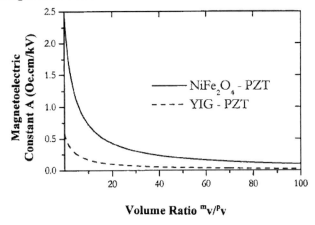

Figure 6.1 Dependence of magnetoelectric constant A on volume fraction of magnetostrictive and piezoelectric phases $^mv/^pv$. The results are for bilayer composites of nickel ferrite–PZT and YIG–PZT.

6.2 BASIC THEORY: MACROSCOPIC HOMOGENEOUS MODEL

Now, let us assume that the composite is a macroscopically homogeneous material, i.e., the ME composite can be considered as an effective media with a ME effect that is not observed separately in each component phase. In this case, the influence of an external electric field E on the magnetic resonance spectrum can be described by the term [1, 20, 21]

$$W = \int_V \left(W_0 + \Delta W_{ME}\right)d^3x$$

in the thermodynamic potential; where W_0 is thermodynamic potential at $\mathbf{E} = 0$, and

$$\Delta W_{ME} = B_{ikn}E_iM_kM_n + b_{ijkn}E_iM_kM_n \tag{6.11}$$

and where B_{ijk} and b_{ijkn} are the linear ME coefficients which are third rank axial tensors. The potential ΔW_{ME} can be found [7] by considering the elastic, magnetostrictive, piezoelectric, and electrostrictive contributions at specified boundary conditions.

First, let us obtain the expression for the magnetic susceptibility tensor of the ferrite phase in the presence of ME interactions. In this case, the composite will be influenced by an applied electric field, and also by the constant and variable magnetic fields that are necessary to observe magnetic resonance. Solving the linearized equation for the magnetization vector rotation, without taking into account losses, and with taking into account the effect of demagnetization factors, the following expression can be obtained [19] for the magnetic susceptibility tensor:

$$\chi^m = \begin{bmatrix} \chi_1 & \chi_5 - i\chi_a & 0 \\ \chi_5 - i\chi_a & \chi_2 & 0 \\ 0 & 0 & 0 \end{bmatrix}, \tag{6.12}$$

where

$$\chi_1 = D^{-1}\gamma^2 M_0 \left[H_{03'} + M_0 \sum_i \left(N_{1'1'}^i - N_{3'3'}^i \right) \right],$$

$$\chi_2 = D^{-1}\gamma^2 M_0 \left[H_{03'} + M_0 \sum_i \left(N_{2'2'}^i - N_{3'3'}^i \right) \right],$$

$$\chi_3 = -D^{-1}\gamma^2 M_0^2 \sum_i N_{1'2'}^i,$$

$$\chi_a = D^{-1}\gamma M_0 \omega,$$

$$D = \omega_0^2 - \omega^2,$$

$$\omega_0^2 = \gamma^2 \left[H_{03'} + \sum_i \left(N_{1'1'}^i - N_{3'3'}^i \right) M_0 \right] \left[H_{03'} + \sum_i \left(N_{2'2'}^i - N_{3'3'}^i \right) M_0 \right] - \left(\sum_i N_{1'2'}^i M_0 \right)^2,$$

and where γ is the magnetomechanical coefficient, ω is circular frequency, $N_{kl}^{i=a}$ are the effective demagnetization factors describing the effective fields of the magnetic anisotropy, $N_{kl}^{i=m}$ is the demagnetization of the sample form; and 1', 2', and 3' is a system of coordinates in which the axis 3' is directed along that of the spontaneous magnetization.

The following additional term with ($i = E$) must be included into the magnetic anisotropy, to account for the ME constants B_{ij}

$$N_{k'n'}^{i'} = 2B_{ikn}E_{oi}\beta_{k'k}\beta_{n'n},$$ (6.13)

where $\hat{\beta}$ is the matrix of direction cosines projected along the axes (1', 2', 3') from the axes of the crystallographic system (1, 2, 3). We must also account for losses in the equation for the magnetization vector rotation, which will take the form of $i\cdot\omega\cdot\alpha\cdot(\mathbf{M_0} \times \mathbf{m})/M_0$: where α is the loss parameter and m *is* the variable magnetization. This results in a complex value for the magnetic susceptibility, given as $\chi = \chi' + i\chi''$, where

$$\chi' = \chi_0 \frac{\omega_0^2(\omega_0^2 - \omega^2 + 2\alpha^2\omega^2)}{(\omega_0^2 - \omega^2)^2 + 4\alpha^2\omega_0^2\omega^2},$$

$$\chi'' = \chi_0 \frac{\alpha\omega\omega_0(\omega_0^2 + \omega^2)}{(\omega_0^2 - \omega^2)^2 + 4\alpha^2\omega_0^2\omega^2}, \quad \chi_0 = \gamma\frac{M_0}{\omega_0}.$$

As an example, we consider a special case when the magnetostrictive component is magnetized in a plane (110) under the corner θ to the cubic axis [001]. Theoretical calculations are essentially simplified by choice of magnetization direction coinciding to direction of stress $^mT_{33}$. Then the additional term of energy can be presented as

$$\Delta W_{ME} = \frac{3}{8M^2}\left(\lambda_{111} - \lambda_{100} + (\lambda_{100} - \lambda_{111})\cos 2\theta\right)M_1^2 \,^mT_{33}$$

$$+ \frac{9}{32M^2}\left(\lambda_{111} - \lambda_{100} + (\lambda_{100} - \lambda_{111})\cos 4\theta\right)M_2^2 \,^mT_{33}$$

$$+ \frac{3}{8M^2}\left((\lambda_{100} - \lambda_{111})\sin 2\theta + \frac{3}{2}(\lambda_{100} - \lambda_{111})\sin 4\theta\right)M_2 M_3 \,^mT_{33}$$

$$+ \frac{3}{8M^2}\left(\frac{-9\lambda_{111} - 7\lambda_{100}}{4}\right) + (\lambda_{111} - \lambda_{100})\cos 2\theta +$$

$$+ \frac{3}{4}(\lambda_{111} - \lambda_{100})\cos 4\theta)M_3^2 \,^mT_{33}.$$

(6.14)

For the special case considered above when $M_0//[111]$, we can obtain

$$\Delta W_{ME} = \left(\frac{\lambda_{111} - \lambda_{100}}{2}\right) M_1^2 \, {}^mT_{33} + \left(\frac{\lambda_{111} - \lambda_{100}}{2}\right) M_2^2 \, {}^mT_{33}$$
$$+ \left(\frac{-\lambda_{111} - \lambda_{100}}{2}\right) M_3^2 \, {}^mT_{33}. \tag{6.15}$$

Using Eq. 6.12 it is then easy to show that the resonance line is shifted under action of an applied electric field, and using a linear approximation for N_{kl}^E, we have

$$\delta H_E = -\frac{M_0}{Q_1}\left[Q_2(N_{11}^E - N_{33}^E) + Q_3(N_{22}^E - N_{33}^E) + Q_4 N_{12}^E\right], \tag{6.16}$$

where

$$Q_1 = 2H_3 + M_0 \sum_{i \neq E}\left[\left(N_{11}^E - N_{33}^E\right) + \left(N_{22}^E - N_{33}^E\right)\right],$$

$$Q_2 = \left[H_3 + M_0 \sum_{i \neq E}\left(N_{22}^E - N_{33}^E\right)\right],$$

$$Q_3 = \left[H_3 + M_0 \sum_{i \neq E}\left(N_{11}^E - N_{33}^E\right)\right],$$

$$Q_4 = 2M_0 \sum_{i \neq E} N_{12}^i.$$

Equation 6.15 allows us to define the ME constants of the composite, and hence, to interpret data concerning the ME effect.

6.2.1 Uniaxial Structure

Let us apply these equations to the consideration of a ME material with a uniaxial structure having 3m symmetry. In this case, Eq. 6.11 for the thermodynamic potential can be written as:

$$\begin{aligned}
\Delta W_{ME} = {} & E_1[B_{11}(M_1^2 - M_2^2) - 2B_{22}M_1M_2 + 2B_{14}M_2M_3 \\
& + 2B_{15}M_1M_3] + E_1[B_{22}(M_2^2 - M_1^2) - 2B_{11}M_1M_2 \\
& + 2B_{15}M_2M_3 - 2B_{14}M_1M_3] + E_3(B_{33} - B_{31})M_3^2 \\
& + E_1^2[(b_{11} - b_{12})M_1^2 + (b_{13} - b_{12})M_3^2 + 2b_{14}M_2M_3 \\
& - 2b_{25}M_1M_3 + 2b_{16}M_1M_2] + E_2^2[(b_{11} - b_{12})M_2^2 \\
& + (b_{13} - b_{12})M_3^2 - 2b_{14}M_2M_3 + 2b_{25}M_1M_3 - 2b_{16}M_1M_2] \\
& + E_3^2(b_{33} - b_{31})M_3^2 + 2E_2E_3(2b_{41}M_1^2 + b_{41}M_3^2 \\
& + 2b_{44}M_2M_3 + 2B_{45}M_1M_3 + 2B_{52}M_1M_2) \\
& + 2E_1E_3(-2b_{52}M_1^2 - b_{52}M_3^2 - 2b_{45}M_2M_3 + 2b_{44}M_1M_3 \\
& + 2B_{41}M_1M_2) + 2E_1E_2(-2b_{16}M_1^2 - b_{16}M_3^2 + 2b_{25}M_2M_3 \\
& + 2b_{14}M_1M_3 + 2b_{66}M_1M_2),
\end{aligned} \tag{6.17}$$

where the usual designations for two indexes is used: $B_{ijk} = B_{i\lambda}$ where $\lambda = 1, 2, 3, 4, 5, 6$ corresponds to $j = k = 1, j = k = 2, j = k = 3, j = 2$ and $k = 3, j = 1$ and $k = 3, j = 1$ and $k = 2$. Using this two indices system, we can then use either tensor (3 × 3 ×3) or matrix (6 × 6) forms for the third rank linear ME coefficient.

Using effective demagnetization factors, the effective magnetic field can be calculated as [19]

$$\bar{H}_E = -\partial W_{ME}/\partial \bar{M} = -N^E \bar{M}. \tag{6.18}$$

Equation 6.17 can be written in a system of coordinates (1', 2', 3') for which the axis 3' coincides with the direction of spontaneous magnetization M_0, as shown in Fig. 6.2. The components of H_k^E can then be given as

$$H_{k'}^E = \beta_{k'k} H_k^E, \tag{6.19}$$

where the direction cosine matrix $\hat{\beta}$ follows from the system of coordinates in Fig. 6.2.

$$\hat{\beta} = \begin{pmatrix} 1 & 0 & 0 \\ 0 & \cos\Theta & -\sin\Theta \\ 0 & \sin\Theta & \cos\Theta \end{pmatrix}. \tag{6.20}$$

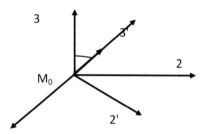

Figure 6.2 System of coordinates for the multilayer composite.

Accordingly, we also have that

$$M_K = \beta_{K'K} M_{K'}. \tag{6.21}$$

By substituting Eq. 6.18 into Eq. 6.19, and by considering Eqs. 6.19–6.21, we can find that

$$N_{11}^E - N_{33}^E = 4(B_{11}E_1 - B_{22}E_2) + 2(b_{11} - b_{12})(E_1^2 - E_2^2)$$
$$+ 8(b_{41}E_2E_3 - b_{16}E_1E_2 - b_{52}E_1E_3) + 2g_2\cos^2Q + 2g_3\sin^2Q,$$

$$N_{22}^E - N_{33}^E = 2g_2\cos^2Q + 4g_3\sin^2Q,$$

$$N_{12}^E = [-2(B_{11}E_2 + B_{22}E_1) + 2b_{16}(E_1^2 - E_2^2)$$
$$+ 4(b_{41}E_1E_3 + b_{52}E_2E_3 + b_{66}E_1E_2)]\cos Q$$
$$+ 2(B_{14}E_2 - B_{15}E_1) + b_{25}(E_1^2 - E_2^2)$$
$$- 4(b_{14}E_1E_2 + b_{44}E_1E_3 + b_{45}E_2E_3)]\sin Q, \tag{6.22}$$

where
$$g_2 = -B_{11}E_1 + B_{22}E_2 + (B_{31} - B_{33})E_3 + (b_{12} - b_{13})E_1^2 + (b_{11} - b_{13})E_2^2$$
$$+ (b_{31} - b_{33})E_3^2 + 2(b_{16}E_1E_2 - b_{41}E_2E_3 + b_{52}E_1E_3),$$
$$g_3 = -B_{14}E_1 - b_{15}E_2 + b_{14}(E_2^2 - E_1^2) + 2(b_{25}E_1E_2 + b_{44}E_2E_3 - b_{45}E_1E_3),$$
$$B_{66} = (b_{11} - b_{12})/2.$$

Please note that indexes in the right-hand side of Eq. 6.22 correspond to those of the crystallographic system of coordinates. We then consider the case when the electric field is directed along the axis of symmetry: i.e., $E_1 = E_2 = 0$, and $E_3 = E$. For this specific orientation, we obtain

$$N_{11}^E - N_{33}^E = 2[(B_{31} - B_{33})E + (b_{31} - b_{33})E^2]\cos^2 Q,$$

$$N_{22}^E - N_{33}^E = 2[(B_{31} - B_{33})E + (b_{31} - b_{33})E^2]\cos^2 Q,$$

$$N_{12}^E = 0. \qquad (6.23)$$

Taking into account the magnetic cubic anisotropy, and the anisotropy form for the ferrite layer oriented along the crystallographic (111) plane (i.e., E'', H'', [111]), it is possible to find that the resonance frequency depends on magnetic and electric fields, as given by

$$\omega/\gamma = H_3 + M_0[4/3 \times K_1/M_0^2 - 4p + 2(B_{31} - B_{33})E + 2(b_{31} - b_{33})E^2], \qquad (6.24)$$

where M_0 is now the effective saturation magnetization of the multilayer composite as an effective magnetic media. If we wish to only consider how the resonance frequency shifts under an applied electric field due to ME interactions, we neglect all other contributions to Eq. 6.24 other than those provided via the linear ME coefficient B_{ijk}, which can be given as

$$\delta H_E = -2M_0(B_{33} - B_{31})E. \qquad (6.25)$$

From Eqs. 6.25 and 6.9, we can estimate the value of the ME coefficient in the microwave range [22]: $2M_0(B_{31} - B_{33}) = 0.1$ Oe cm/kV for YIG–PZT composite, $2M_0(B_{31} - B_{33}) = 0.6$ Oe cm/kV for a lithium ferrite (LFO)–PZT composite, and $2M_0(B_{31} - B_{33}) = 1.4$ Oe cm/kV for a NFO–PZT composite. Similar analyses can be performed to determine the ME constants for other orientations of E and H fields.

Using the above values of the ME constant in the microwave range, it is now possible to define the changes in the magnetic susceptibility under an external electric field by Eq. 6.12. Results of such calculations of the magnetic susceptibility for bilayer composites are given in Figs. 6.3, 6.4, and 6.13 [22, 27]. In Fig. 6.13, the dependence of the imaginary part of the magnetic susceptibility on dc bias is shown for a bilayer disc composite, where both constant magnetic and electric fields are perpendicular to the plane of the sample. For $E = 0$, a typical resonant line is observed. External constant electric field simply results in a shift of the resonant magnetic field, where the magnitude of the shift is proportional to the ME constants, which in turn are defined by the piezoelectric and magnetoelastic constants. Prior investigations have shown shifts of 330 Oe for NFO–PZT, and 22 Oe for YIG–PZT. The large ME effect in NFO is due to its much larger magnetostriction, in comparison to YIG. Clearly, the theoretical model presented above is an important tool for predicting ME interaction in the microwave frequency range. It allows us to define the ME constants, and their dependence on crystallography, material systems, and boundary parameters.

Figures 6.3 and 6.4 illustrate the dependence of the resonant (9.2 GHz) magnetic susceptibility on an external electric field. The width of the resonant line in terms of electric field is inversely proportional to the value of $2M_0 (B_{31} - B_{31})$. From Eq. 6.12, it then follows that a narrow resonant line is characteristic of a composite with strong ME effects. Accordingly, a bilayer composite of ferrite nickel–PZT has a narrower resonant line, relative to a similar composite structure with YIG–PZT.

The magnetic resonance data shown above taken under sweeping magnetic and electric fields is of possible interests for applications in solid state electronics. Electric field tunability of the magnetic subsystem allows for simplified methods for frequency tuning of microwave signals. In consideration that the dielectric breakdown strength of insulating thin layers that are well prepared can reach values of up to 300kV/cm, it is possible to conceive of a ME resonator based on a simple bilayer composite structure of nicel ferrite–PZT that can be tuned by up to 925 MHz, and by 60 MHz for a similar structure of YIG–PZT.

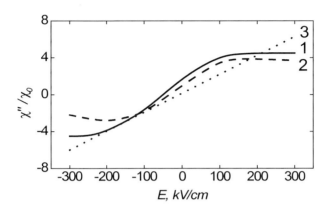

Figure 6.3 Dependence of real part of magnetic susceptibility for layered structure at frequency 9.3 GHz on the electric field: 1 — lithium ferrite–PZT, 2 — nickel ferrite–PZT, 3 — YIG–PZT.

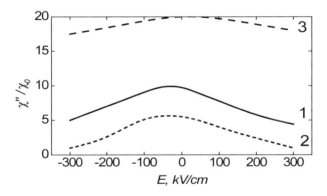

Figure 6.4 Dependence imaginary part of magnetic susceptibility for layered structure at frequency 9.3 GHz on the electric field: 1 — lithium ferrite–PZT, 2 — nickel ferrite–PZT, 3 — YIG–PZT.

6.3 LAYERED COMPOSITE WITH SINGLE CRYSTAL COMPONENTS

ME interaction in ferromagnetic–ferroelectric heterostructures is mediated by elastic strain. Under an external electric field E, the piezoelectric component phase strains. This strain is then transferred to the ferrite component phase, resulting in a shift of its resonant magnetic field. Modeling of the microwave ME effect in layered or bulk composites has revealed two important types of ME interaction [22, 23], which are as follows: (i) between microwave magnetic and microwave electric fields and (ii) between microwave magnetic and constant electric fields.

There are practical difficulties in observing the bias dependence of the magnetic susceptibility shown earlier in Fig. 6.13. Measurements on samples of bulk composites of 90%YIG-10%PZT [9] have shown only weak ME interactions, due to a low volume fraction of PZT; but, when the composite was loaded with higher volume fractions of PZT, a FMR line broadening masked the effect of an external electric field. Such line broadening, however, can easily be eliminated in layered structure by using single crystal films of YIG.

Next, let us discuss prior experimental results of microwave ME interactions in bilayer structures of YIG and lead magnesium niobate–lead titanate (PMN–PT) single crystals, and compare data to theoretical predictions [22, 23]. Single crystal heterostructures are preferable for experimental measurements because: (i) a smaller FMR line width facilitates exact definition of the resonant magnetic field shift, and accordingly the ME constants; and (ii) theoretically, the ME constant of single crystals should be larger relative to a corresponding polycrystalline sample [16, 22]. Bilayer structures (4×4 mm^2) were made using epitaxial YIG films of thickness 1–110 μm grown on (111) gallium–gadolinium garnet (GGG) substrates that were 0.5 mm thick.

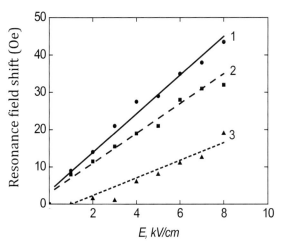

Figure 6.5 Magnetoelectric effect in two-layer structures YIG (111) on GGG and PMN–PT (100) at frequency 9.3 GHz. Static fields E and H are parallel to an axis YIG [111] and perpendicular to sample plane. Shift of resonant magnetic field is shown as function E for different thickness of YIG film: 1 — 4.9 μm, 2 — 58 μm, 3 — 110 μm.

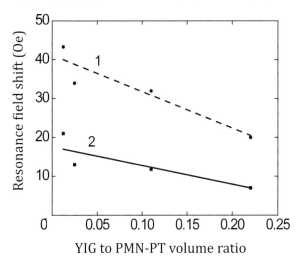

Figure 6.6 Measured shift of resonant magnetic field in electric field $E = 8$ kV/cm as a function of YIG-to-PMN-PT volume ratio for out-of-plane and in-plane H: 1 — $H//[111]$, 2 — $H//(111)$."

Measurements of the ME effect were then performed using an EPR spectrometer at a frequency 9.3 GHz [22, 23]. The input power was 0.1 mW, which corresponds to a magnetic field of about 130 A/м. The ME composite was placed outside of the resonator to eliminate potential overload of the resonator at the resonant absorption peak. Measurements were made under an electric field ($0 < E < 8$ kV/cm) applied perpendicular to the plane of composite for the following directions of magnetization: (i) H'', [011], and (ii) $H//$,(111).

For $E = 0$, the FMR line width was 0.5–4 Oe, depending on film thickness. Under external electric fields, the resonant magnetic field decreased with increasing E, as shown in Fig. 6.13. The dependence of resonant magnetic field shift on external dc electric field is shown in Fig. 6.5: data are given for $H//$, $E//$, [111] for different thickness of YIG. A linear dependence of the resonant magnetic field shift on dc electric field can clearly be seen in this figure. The ME constant $A = dH_E/E$ can be estimated from data to be 5.4 Oe cm/kV for the composite with a YIG thickness of 4.9 mm. Increase of the YIG thickness to 58 μm resulted in a reduction of the ME constant to 4.4 Oe cm/kV.

Measurements of the resonant magnetic field shift were also performed for $H//$, (011) plane of YIG. Along this direction, the resonant magnetic field shift is not as large as that when $H//$, (111). Accordingly, the values of dH_E and A were smaller for $H//$, (011). Prior investigations have been performed on magnetostrictive YIG and piezoelectric PMN–xPT bilayer composites as a function of the relative volume fractions. Figure 6.6 shows that the microwave ME effect depends near linearly on the relative volume fraction. The shift in the resonant magnetic field induced by an external electric field can be seen to decrease with increasing volume fraction of YIG, consistent with the predictions of the last section.

For better comparisons of the experimental data with theoretical predictions, we should consider the additional term for the thermodynamic potential given by Eq. 4.11, which describes influence of external constant electric field E on the magnetic resonance spectrum. Let us then consider a bilayer composite of YIG and PMN–PT. First, we should calculate the deformation induced in PMN–PT by application of E. The generalized Hooke's law in this case, including piezoelectric constitutive relations, can be given as:

$$
\begin{cases}
{}^{P}S_3 = {}^{P}d_{33}E + {}^{P}s_{13}({}^{P}T_1 + {}^{P}T_2) + {}^{P}S_{33}\,{}^{P}T_2, \\[2mm]
{}^{P}S_1 = {}^{P}d_{31}E + {}^{P}s_{11}\,{}^{P}T_1 + {}^{P}s_{12}\,{}^{P}T_2 + {}^{P}s_{13}\,{}^{P}T_3, \\[2mm]
{}^{P}S_2 = {}^{P}d_{31}E + {}^{P}s_{12}\,{}^{P}T_1 + {}^{P}s_{11}\,{}^{P}T_2 + {}^{P}s_{13}\,{}^{P}T_3.
\end{cases}
\tag{6.26}
$$

Now, assume that the direction of the spontaneous polarization in the piezoelectric phase coincides with the [111] axis of the ferrite phase. In this case, the compliance tensor of the ferrite phase is

$$
{}^{m}s_{ijkl} = {}^{m}s_{i'j'k'l'}\,\beta_{ii'}\,\beta_{jj'}\,\beta_{kk'}\,\beta_{kk'},
\tag{6.27}
$$

where (1, 2, 3) is a system of coordinates in which the axis 3 is directed along direction of the spontaneous magnetization, and β is the direction of cosine matrix. For the ferrite phase we can then determine that

$$
\begin{cases}
{}^{m}S_3 = {}^{m}s_{13}({}^{m}T_1 + {}^{m}T_2) + {}^{m}s_{33}\,{}^{m}T_3, \\[2mm]
{}^{m}S_1 = {}^{m}s_{11}\,{}^{m}T_1 + {}^{m}s_{12}\,{}^{m}T_2 + {}^{m}s_{13}\,{}^{m}T_3, \\[2mm]
{}^{m}S_2 = {}^{m}s_{12}\,{}^{m}T_1 + {}^{m}s_{11}\,{}^{m}T_2 + {}^{m}s_{13}\,{}^{m}T_3.
\end{cases}
\tag{6.28}
$$

Next, let us use the usual boundary conditions for a mechanically free composite

$$
{}^{P}S_1 = {}^{m}S_1,\ {}^{P}S_2 = {}^{m}S_2,\ {}^{P}T_3 = {}^{m}T_3 = 0,
$$
$$
{}^{P}T_1 = -{}^{m}v/{}^{P}v \times {}^{m}T_1,\ {}^{P}T_2 = -{}^{m}v/{}^{P}v \times {}^{m}T_2,
\tag{6.29}
$$

where ${}^{m}v$ and ${}^{P}v$ are the volume fractions of the ferrite and piezoelectric phases, respectively. Solutions of Eqs. 6.26 and 6.28 by taking into account Eq. 6.29 gives for the stress on the magnetostrictive phase

$$
{}^{m}T_1 = {}^{m}T_2 = -{}^{P}d_{31}\,E\,{}^{P}v/[\,{}^{P}v({}^{m}s_{11} - {}^{m}s_{12}) + {}^{m}v({}^{P}s_{11} - {}^{P}s_{12})].
\tag{6.30}
$$

The shift of the resonant magnetic field (which depends on the volume fractions of magnetic and piezoelectric phases) can then be calculated from Eq. 6.15 by taking into account Eq. 6.30 for the following two cases in the (111) plane of YIG: (i) H'', [111], and (ii) H'', [011].

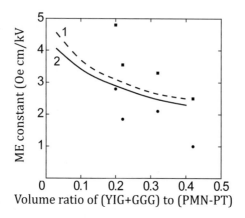

Figure 6.7 Comparison of calculation (line) and data of ME constant for magnetic field laying in sample plane (circles) and perpendicular to sample plane (squares). Values of ME constant are presented as function of (YIG + GGG) and PMN–PT volume fractions.

Figure 6.7 illustrates the dependence of the ME constant A (which is numerically equal to the shift of the resonant line under $E = 1$ kV/cm) for a by-layer composite of YIG + GGG and PMN–PT. The thicknesses of the YIG + GGG and PMN–PT layers were 0.5 mm and 0.1 mm, respectively. Calculations were performed using the following material parameters: $^{p}d_{31} = -600 \times 10^{-12}$ m/V, $^{p}d_{33} = 1500 \times 10^{-12}$ m/V, $^{p}s_{11} = 23 \times 10^{-12}$ m²/N, $^{p}s_{12} = -8.3 \times 10^{-12}$ m²/N, $\lambda_{100} = -1.4 \times 10^{6}$, $\lambda_{111} = -2.4 \times 10^{-6}$, $4\pi M_{0} = 1750$ G, $H_{a} = -42$ Oe, $^{m}s_{11} = 4.8 \times 10^{-12}$ m²/N, and $^{m}s_{12} = -1.4 \times 10^{-12}$ m²/N. Figure 6.7 shows (i) that the ME constant decreases with increasing YIG volume fraction; and (ii) that the ME interaction is larger for magnetic fields perpendicular to composite plane than for those parallel to it. In this figure, good qualitative and quantitative agreement can be seen between theory and data.

From the above results, we can conclude to obtain an optimum ME effect: (i) that the volume fraction of the piezoelectric phase should be high; (ii) that it is necessary to use a piezoelectric component phase with a large piezoelectric coefficient; and (iii) that it is necessary to use a magnetostrictive component phase with a small saturation magnetization and a high magnetostriction.

6.4 RESONANCE LINE SHIFT BY ELECTRIC SIGNAL WITH ELECTROMECHANICAL RESONANCE FREQUENCY

As we discussed in the last chapter, the electromechanical resonance (EMR) frequency range has a significant enhancement of the ME interaction. Here, we estimate the influence of an external electric field at the EMR frequency on the magnetic resonance spectrum for layered composite with a magnetostrictive ferrite component phase [24, 25].

To determine the magnetic resonance line shift, we use Eq. 6.16. The calculation of the mechanical stress on the ferrite phase imposed by the piezoelectric on driven in the EMR range can be obtained from the equation of motion in Eq. 3.1, which has a general solutions of form given in Eqs. 3.45 and 3.48. The integration constants can be obtained by assuming that the composite is mechanically free. In this case, the stress in the ferrite component phase can be given as

$$ {}^{m}T_1 = \frac{[\sin(kx)tg(kL/2)+\cos(kx)]E_3 \, {}^{pd}_{31}v}{{}^{p}s_{11}(1-v)+ {}^{m}s_{11}v}. \tag{6.31}$$

From Eq. 6.16 by including Eq. 6.31, it is possible to find out how the ME constant depends on the relative phase volume fractions. The results are shown in Fig. 6.8. In addition, the dependence of the magnetic resonance line shift on external electrical field is illustrated in Fig. 6.9.

In these figures, it can be seen to achieve an equivalent resonant line shift under external electric field at the EMR frequency that a field of only 1/100th of that of a dc electric bias needs to be applied. These results show that the microwave ME properties should also be dramatically enhanced by driving the composite at its EMR frequency.

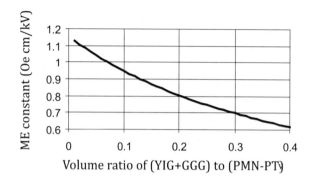

Figure 6.8 Calculated dependence of ME constant in the EMR range for magnetic field perpendicular to sample plane on (YIG + GGG) and PMN–PT volume fractions.

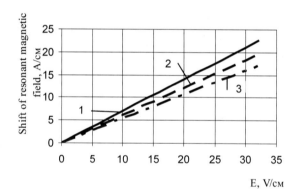

Figure 6.9 Calculated shift of resonant magnetic field in external electric field with EMR frequency for bias fields perpendicular to sample plan: 1 — thickness of YIG film is 4.9 µm, 2 — 58 µm, 3 — 110 µm.

6.5 ME EFFECT AT MAGNETOACOUSTIC RESONANCE RANGE

Now, let us consider the bilayer composite of Fig. 6.10, which is for ferrite and piezoelectric single crystal layers, where the ferrite one is in a saturated single domain state [25]. This state has two important advantages. First, when domains are present, the acoustic losses in the high-frequency range are significant. Second, the large magnetic susceptibility relaxes out with increasing frequency, and saturated single domain states are the only way that respectable effective susceptibilities can be achieved at FMR [26].

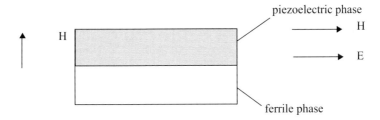

Figure 6.10 Two-layer structure on the basis of single crystal phases.

The free energy density of a single crystal ferrite phase can be given as

$$^{m}W = W_{H} + W_{an} + W_{ma} + W_{ac},\qquad(6.32)$$

where $W_{H} = -\mathbf{M}\cdot\mathbf{H}$ is Zeeman's energy, $W_{an} = K_{1}/M^{4}(M_{1}^{2}M_{2}^{2} + M_{2}^{2}M_{3}^{2} + M_{3}^{2}M_{1}^{2})$ is cubic anisotropy energy, K_{1} is a constant of the cubic anisotropy, $W_{ma} = B_{1}/M^{2^{\blacksquare}}(M_{1}^{2}\,^{m}S_{1} + M_{2}^{2}\,^{m}S_{2} + M_{3}^{2}\,^{m}S_{3}) + B_{2}/M^{2}(M_{1}M_{2}\,^{m}S_{6} + M_{2}M_{3}\,^{m}S_{4} + M_{1}M_{3}\,^{m}S_{5})$ is the magnetoelastic energy, B_{1} and B_{2} are magnetoelastic constants, and $W_{ac} = ½\,^{m}c_{11}(^{m}S_{1}^{2} + ^{m}S_{2}^{2} + ^{m}S_{3}^{2}) + ½\,^{m}c_{44}(^{m}S_{4}^{2} + ^{m}S_{5}^{2} + ^{m}S_{6}^{2}) + ^{m}c_{12}(^{m}S_{1}\,^{m}S_{2} + ^{m}S_{2}\,^{m}S_{3} + ^{m}S_{1}\,^{m}S_{3})$ is the elastic energy. In Eq. 6.32, it is supposed that the material is uniformly magnetized. On the basis of a generalized Hooke's law, we can write the stresses of the piezoelectric component phase as

$$^{p}T_{4} = ^{p}c_{44}\,^{p}S_{4} - e_{p15}\,^{p}E_{2},$$

$$^{p}T_{5} = ^{p}c_{44}\,^{p}S_{5} - e_{p15}\,^{p}E_{1},\qquad(6.33)$$

where e_{p15} is a shear piezoelectric coefficient. The equations of motion for the ferrite and piezoelectric phases are then

$$\partial^2(^mu_1)/\partial t^2 = \partial^2(^mW)/(\partial \times \partial^mS_1) + \partial^2(^mW)/(\partial y\partial^mS_6)$$
$$+ \partial^2(^mW)/(\partial z\partial^mS_5),$$

$$\partial^2(^Pu_1)/\partial t^2 = \partial(^mT_1)/\partial\times + \partial(^mT_1)/\partial y + \partial(^mT_1)/\partial z,$$

$$\partial^2(^Pu_2)/\partial t^2 = \partial(^mT_2)/\partial\times + \partial(^mT_2)/\partial y + \partial(^mT_2)/\partial z.$$

$$(6.34)$$

The equation of the magnetization vector rotation are given by

$$\partial M/\partial t = -\gamma\,[M, H_{\text{eff}}], \tag{6.35}$$

where $H_{\text{eff}} = -\partial\,(^mW)/\partial\,M$.

To simply calculations, we assume waves with the circular polarization

$$m^+ = m_1 + i\,m_2,$$

$$H^+ = H_1 + i\,H_2,$$

$$E^+ = E_1 + i\,E_2,$$

$$u^+ = u_1 + i\,u_2, \tag{6.36}$$

where m is a variable magnetization, and u is the displacement. By taking into account Eq. 6.35, Eqs. 6.33 and 6.34 take the form of

$$\omega\,m^+ = \gamma\,(H_0\,m^+ - 4\,\pi\,M_0\,m^+ - M_0\,H^+),$$

$$-\omega^2\,{}^m\rho\,{}^mu^+ = {}^mc_{44}\,\partial^2(^mu^+)/\partial z^2,$$

$$-\omega^2\,{}^P\rho\,{}^Pu^+ = {}^Pc_{44}\,\partial^2(^Pu^+)/\partial z^2. \tag{6.37}$$

The boundary conditions at the interface between layers and at the top/bottom planes of the composite are given by

$${}^mu^+ = {}^Pu^+ \text{ at } z = 0,$$

$${}^mc_{44}\,\partial(^mu^+)/\partial z + B_2\,m^+/M_0 = 0 \text{ at } z = {}^mL,$$

$$^{m}c_{44}\,\partial(^{m}u^{+})/\partial z + B_{2}\,m^{+}/M_{0} = {^{p}c_{44}}\,\partial(^{p}u^{+})/\partial z - {^{p}e_{15}}\,^{p}E^{+} \text{ at } z = 0,$$

$$^{p}c_{44}\,\partial(^{p}u^{+})/\partial z - {^{p}e_{15}}\,^{p}E^{+} = 0 \text{ at } z = -{^{p}L}, \tag{6.38}$$

where ^{m}L and ^{p}L are the thicknesses of the ferrite and piezoelectric layers, respectively. Finally, the electric field E induced in the piezoelectric component can be defined by the open electric circuit condition

$$D^{+} = \frac{1}{^{p}L}\int_{-^{p}L}^{0} {^{p}D^{+}}\,dz = 0, \tag{6.39}$$

where $^{p}D^{+} = {^{p}e_{15}}\,^{p}S^{+} + {^{p}\varepsilon_{11}}\,^{p}E^{+}$ is the dielectric displacement of the piezoelectric layer.

Substitution of the solution for Eq. 6.37 into Eq. 6.39, and by taking into account Eq. 6.38, we can derive an expression for the ME voltage coefficient:

$$\begin{aligned}
|E^{+}/H^{+}| = {}& \gamma\,B_{2}\,{^{p}c_{44}}\,k_{p}\,{^{p}e_{15}}\,[1 - \cos(k_{p}\,{^{p}L})]^{2}/\{(\omega - \gamma H_{0} + 4\pi\,\gamma M_{0}) \\
& \times [-\tfrac{1}{2}\,{^{p}c_{44}}\,{^{p}\varepsilon_{33}}\,k_{p}\,{^{p}L}\,\sin(2k_{p}\,{^{p}L})\,({^{p}c_{44}}k_{p} + {^{m}c_{44}}k_{m}) \\
& + (\cos(k_{p}\,{^{p}L}) - 1) \times {^{p}e_{15}}^{2}[(\cos(k_{p}\,{^{p}L})({^{m}c_{44}}k_{m} + 2\,{^{p}c_{44}}k_{p}) \\
& + {^{m}c_{44}}k_{m}]]\}, \tag{6.40}
\end{aligned}$$

where

$$k_{m} = \omega\sqrt{\frac{^{m}\rho}{^{m}c_{44}}}, \quad k_{p} = \omega\sqrt{\frac{^{p}\rho}{^{p}c_{44}}}$$

As follows from Eq. 6.38, there is a connection in the ferrite phase between the displacement and a homogeneous magnetization precession, through boundary conditions on the plate surfaces. Expression 6.40 shows that if the frequency of an applied magnetic field equals that of the magnetization precession ($\omega_{0} = \gamma H_{0} - 4\pi\gamma M_{0}$), then the value of the ME voltage coefficient will be increased significantly. This enhancement is due to a coupling between the strain induced by an applied magnetic field in the range of the magnetic resonance and a corresponding one in the piezoelectric phase induced by an applied electric field.

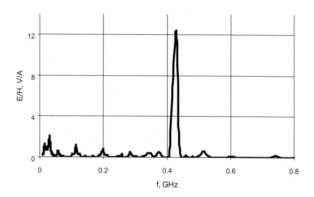

Figure 6.11 Frequency dependence of ME voltage coefficient for L1 = L2 = 0.05 mm, H_0 = 1.9 кОе.

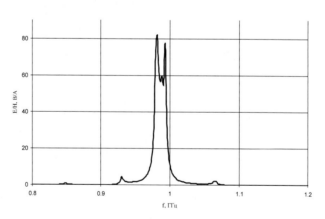

Figure 6.12 Frequency dependence of ME voltage coefficient for L1 = L2 = 0.01 mm, H_0 = 2.1 кЕ.

Figures 6.11 and 6.12 show the dependences of the ME voltage coefficient for a bilayer composite of single crystal YIG and PMN–PT, both of which were calculated by Eq. 6.40. Calculations were performed by introducing a complex frequency to account for magnetoacoustic loss. The real component of the frequency was taken as $\omega_r = 10^7$ rad/s.

For frequencies less than that of the homogeneous magnetization precession, the microwave ME voltage coefficient has a maximum when driven by an electric field in the frequency range of the EMR. Essential to this increase in the microwave ME voltage coefficient is that the EMR frequency of the electric be equal to that for a uniform magnetization precession in the ferrite phase. Even after including magnetoacoustic loss factors (as mentioned above), the microwave ME coefficient can reach giant values of up to 64 V/(cmOe).

Thus, one should use bilayer composites of ferrite and piezoelectric single crystal layers. Such radiocomponents work on the basis of generation of a magnetoelastic waves, or by exciting magnetic resonance by external electric field.

6.6 MICROWAVE AND MM-WAVE ME INTERACTIONS AND DEVICES

6.6.1 Introduction

Ferrite–ferroelectric layered structures are of interest for studies on the fundamentals of high-frequency ME interaction and for device technologies. The electromagnetic coupling in such systems is mediated by mechanical stress: magnetostriction-induced mechanical deformation and piezoelectric effect-induced electric fields [31–43]. Such composites are promising candidates for a new class of dual electric and magnetic field tunable devices based on ME interactions [43–50]. Two types of structures have been studied so far:

  i. *ME coupling in bonded ferrite–piezoelectrics:* An electric field E applied to the composite produces a mechanical deformation in the piezoelectric phase that in turn is coupled to the ferrite, resulting in a shift in the FMR field [16, 22, 42]. The strength of the interactions is measured from the FMR shifts.

ii. ME interactions in unbound ferrite–ferroelectrics: This is a proximity effect in which hybrid spin-electromagnetic waves are formed. During the past several years we have made a substantial progress on measurements and theory of ME coupling and hybrid waves. Strong ME interactions were measured in both cases.

Key accomplishments include the following.

- Resonance ME effects in ferrite–piezoelectric bilayers, at FMR for the ferrite. The ME coupling was measured from data on FMR shifts in an applied electric field E. Low-loss YIG was used for the ferromagnetic phase. Single crystal PMN–PT and PZT were used for the ferroelectric phase.
- Theory and measurements on hybrid spin-electromagnetic waves that originate from ME interactions in layered YIG–BST (bismuth strontium titanate) structures, and theory of microelectronic hybrid wave resonators.
- Design, fabrication, and analysis of composite based devices, including resonators and phase shifters. The unique for such devices is the tunability with E.

There have been many breakthroughs in both experiments and theoretical understanding of the phenomenon.

Our studies on YIG–PZT composites also resulted in the design and characterization of a new class of microwave signal processing devices including resonators, filters, phase shifters, and delay lines for use at 1–10 GHz. The unique and novel feature in ME microwave devices is the tunability with an electric field. The traditional "*magnetic*" tuning in ferrite devices is relatively slow and is associated with large power consumption. The "*electrical*" tuning is possible for the composite and is much faster and has practically zero power consumption.

The efforts have also been extended to millimeter and sub-millimeter wave ME interactions in hexagonal ferrites–ferroelectric layered structures. For the frequency range 20–67 GHz, one could use bilayers consisting of single crystal **M**-type barium or strontium ferrites (Ba**M**, Sr**M**), and **Z**- and **Y**-type ferrites (Co$_2$**Z**, Zn$_2$**Y**, etc.) for the magnetic phase and PZT, PMN–PT, or BST for the ferroelectric phase.

6.6.2 Microwave ME Effects in Ferrite–Piezoelectrics: Theory and Experiment

In the microwave region of the electromagnetic spectrum, the ME effect can be observed in the form of a shift in FMR profile in E (Fig. 6.13). Figure 6.14 shows the data on the resonance field shift as a function of E for a bilayer of epitaxial YIG film and PMN-PT single crystal. Theoretical FMR profiles based on our model are shown in Fig. 6.13 for bilayers with YIG, NFO, or LFO and PZT. For $E = 300$ kV/cm, a shift in the resonance field δH_E that varies from a minimum of 22 Oe for YIG/PZT to a maximum of 330 Oe for NFO–PZT is predicted [16, 22]. The strength of ME interactions $A = \delta H_E/E$ is determined by piezoelectric coupling and magnetostriction.

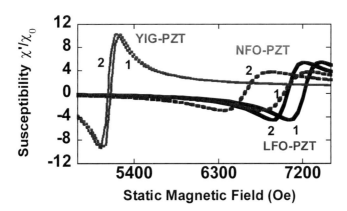

Figure 6.13 Theoretical FMR profiles in ferrite–PZT bilayers at 9.3 GHz for $E = 0$ (labeled — 1) and $E = 300$ kV/cm(−2). E and H are perpendicular to the bilayer [22]. See also Color Insert.

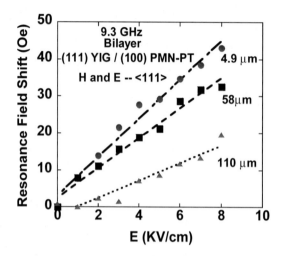

Figure 6.14 Data on FMR field shift vs. *E* for a YIG/PMN–PT bilayer. See also Color Insert.

Figure 6.14 shows our data on δH_z vs. *E* at 9.3 GHz for a bilayer of YIG and PMN–PT and the ME effect is an order of magnitude stronger than predictions in Fig. 6.13 for YIG/PZT. The data also provide evidence for electric field tunability of YIG microwave devices.

6.6.3 Hybrid Spin-Electromagnetic Waves in Ferrite–Dielectrics: Theory and Experiment

The ME effect discussed above takes place in bilayered ferrite–ferroelectric structures when the layers are *tightly bound*, i.e., when the mechanical stress created in one layer is transferred to the neighboring layer. There are, however, other ME phenomena that do not require bonding between the layers and take place simply due to the proximity of two material having different dielectric and magnetic properties [50]. An example of such a phenomenon is the formation of hybrid spin-electromagnetic waves in the layered structures. The nature of the hybrid waves were studied in a YIG–BST structure shown in Fig. 6.15. The dimensions for the YIG and BST were chosen to be equal to maximize the electromagnetic coupling between FMR in YIG and dielectric resonance in BST. Both *H*- and

E-dependence of the hybrid excitations, due to variations in μ and ε, were measured (Fig. 6.16) and compared with theory [50]. The key result is the tunability of the mode by 100 MHz for nominal E and would facilitate the fabrication of E-tunable YIG–BST resonators for microwave circuits.

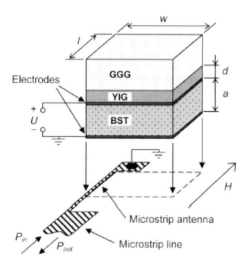

Figure 6.15 Diagram showing the schematics of a YIG–BST layered system for hybrid wave generation [50]. See also Color Insert.

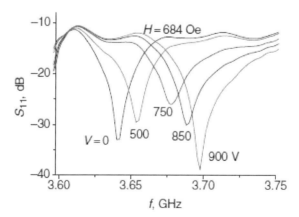

Figure 6.16 Data showing the dependence of the resonant frequency of a YIG–BST resonator on applied voltage across the BST slab [50].

6.6.4 Electric Field Tunable Microwave Devices: YIG–PZT and YIG–BST Resonators

The above studies on microwave ME effects in YIG–PZT, YIG–PMNPT, and YIG–BST led to the design, fabrication, and characterization of a new family of novel signal processing devices that are tunable by both magnetic and electric fields. The device studied included YIG–PZT and YIG–BST resonators, filters, and phase shifters [47–50].

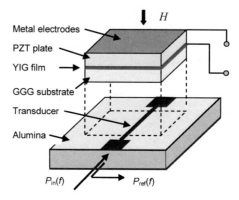

Figure 6.17 YIG–PZT resonator [47]. See also Color Insert.

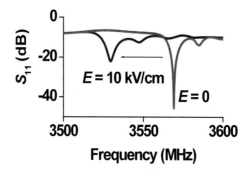

Figure 6.18 Electric field tunable 13 μm YIG–PMN–PT resonator with perpendicular magnetization (H = 3010 Oe) [47]. See also Color Insert.

We have extended the studies described in Sections 6.2 and 6.3 to include electric field effects in YIG resonators [47–50]. Bilayers were made by bonding epitaxial YIG films and single crystal PMN–PT or PZT. Stripline resonators (Fig. 6.17) were subjected to perpendicular or parallel *H*. Figure 6.18 shows the resonator response for *an electric field across PMN–PT*: a shift δf_E, positive or negative depending on the *H* direction, is seen. The tuning obtained here, δH_E = 50 MHz for *E* = 10 kV/cm, is 10 times the width of FMR in YIG and is suitable for device applications. The resonator *Q* ranged from 1000 to 2000. With proper choice of the ferrite and piezoelectric phases and *E*-value, a tuning range of 0.5–1 GHz is quite possible.

Similarly, YIG–BST hybrid wave resonators showed *Q* = 1000 and electric field tunability on the order of 0.1% of the operating frequency [50].

6.6.5 Filters

Efforts on filters focused on stripline and slot-line ferrite–ferroelectric band-pass filters. Non-reciprocal filter may be realized using slot line. Design of our low-frequency ME filter is shown in Fig. 6.19 and representative data on electric field tuning are shown in Fig. 6.20 [52]. The single-cavity ME filter consists of a dielectric ground plane, input and output microstrips, and an YIG–PZT ME-element. Power is coupled from input to output under FMR in the ME element. The frequency dependence of the insertion are shown in Fig. 6.20. A frequency shift of 120 MHz for *E* = 3 kV/cm corresponds to 2% of the central frequency of the filter and is a factor of 40 higher than the line width for pure YIG. from *E* is coupled to the ferrite, leading to a shift δH_z in the FMR as in Fig. 6.14. Theoretical FMR profiles based on our model are shown in Fig. 6.13 for bilayers with YIG, NFO, or LFO and PZT. For *E* = 300 kV/cm, a shift in the resonance field δH_E that varies from a minimum of 22 Oe for YIG/PZT to a maximum of 330 Oe for NFO–PZT is predicted [16, 22]. The strength of ME interactions $A = \delta H_E/E$ is determined by piezoelectric coupling and magnetostriction.

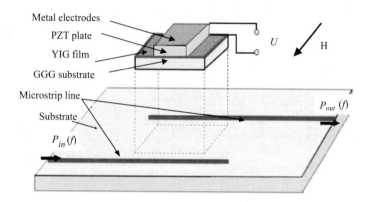

Figure 6.19 A magnetoelectric (ME) band-pass filter. The ME resonator consisted of a 110 μm thick (111) YIG on GGG bonded to PZT [52]. See also Color Insert.

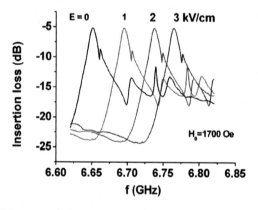

Figure 6.20 Loss vs. f characteristics for a series of E for the YIG–PZT filter [52]. See also Color Insert.

6.6.6 Phase Shifters

We designed (Fig. 6.21) and characterized (Fig. 6.22) a voltage tunable YIG–PZT phase shifter based on FMR. The electric field control of the phase shift $\delta\varphi$ arises through ME interactions [51]. The piezoelectric deformation in PZT in an electric field E leads to a phase shift. For $E = 5$–8 kV/cm applied across PZT, differential-phase shift $\delta\varphi = 90$–$180°$

and an insertion loss of 1.5–4 dB are obtained. Theoretical estimates of $\delta\varphi$ are in excellent agreement with the data.

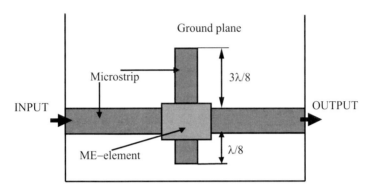

Figure 6.21 Diagram showing the schematics of a YIG/PZT phase shifter [51]. See also Color Insert.

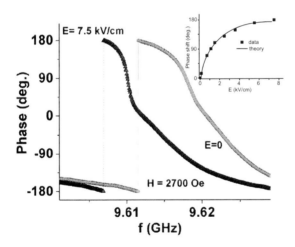

Figure 6.22 The phase angle ϕ vs. frequency f at 9.6 GHz for $E = 0$ and 7.5 kV/cm [51]. See also Color Insert.

It is clear from the discussions here that ME interactions are very strong in the microwave region in bound and unbound ferrite–

ferroelectric bilayers and that a family of dual electric and magnetic field tunable ferrite–ferroelectric resonators, filters, and phase shifters can be realized. The electric field tunability, in particular, is 0.1% or more of the operating frequency of filters and resonators. A substantial differential-phase shift can also be achieved for nominal electric fields.

6.6.7 MM-Wave ME Effects in Bound Layered Structures

In the mm-wave region of the electromagnetic spectrum, the ME effect can be observed in the form of FMR or magnetostatic mode shift. Mechanical stress in the ferroelectric arising from external electric field is coupled to the ferrite, leading to a shift in the resonance field. The objective is to determine the ME constants from FMR data and its dependence on crystallographic, material, and interface parameters.

Preliminary studies are underway on mm-wave ME interactions in BaM–PZT [53]. For BaM with anisotropy field $H_a = 17$ kOe and saturation magnetization $4\pi M = 4.8$ kG, FMR is expected in the frequency range 47–60 GHz for external fields $H_0 = 5$–9 kOe. The condition for FMR, for H_0 perpendicular to the plane of a BaM disk and along the c-axis, is given by

$$\omega_0 = \gamma(H_0 + H_A - 4\pi M),\tag{6.41}$$

where $\gamma = 2.8$ GHz/kOe is the gyrmagnetic ratio. For saturated state when $H_0 > 4\pi M$, $f_0 > 47$ GHz. For an electric field E applied to ferrite/piezoelectric bilayer, the piezoelectric deformation is equivalent to the effect of magnetostriction on FMR, leading to an internal magnetic field $\delta H = AE$, where A is the ME coupling constant. According to our model [16, 41], ME effects at FMR can be taken into account by

$$\omega_0 = \gamma(H_0 + H_A - 4\pi M + AE).\tag{6.42}$$

The measurement, therefore, involve the shift in FMR profiles with and without E.

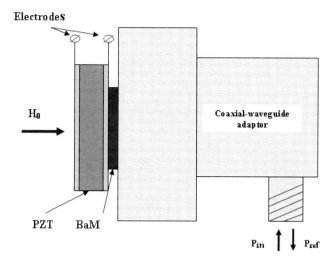

Figure 6.23 Measurement cell for mm-wave ME effects in BaM–PZT bilayers [53]. See also Color Insert.

Our studies were performed on bilayers of single crystal Ba**M** prepared by floating zone method and polycrystalline PZT. The measurement cell is shown in Fig. 6.23 and consists of a coaxial-to-(WR-17) wave guide adapter. The ferrite was bonded to PZT with a fast dry epoxy. The silver electrode at the Ba**M**–PZT interface acted as a classical short load of the waveguide. Studies were performed with E and H_0 perpendicular to the bilayer plane and parallel to the magnetic easy axis of Ba**M**. A vector network analyzer (Agilent-PNA-E8361A) was used for data on scattering matrix element S_{11} vs. f for $E = 0$–10 kV/cm. Figure 6.24 shows representative data on S_{11} vs. f. The absorption peaks for $H_0 = 5980$ Oe and $E = 0$ are identified as electromagnetic modes in Ba**M**. Magnetostatic forward volume waves in BaM are hybridized with dielectric modes, with wave numbers $k_\perp = \pi\sqrt{[(m/a)^2 + (n/b)^2]}$, $m,n = 1,2,...$, where a and b are the in-plane sample dimensions. The absorption peaks labeled 1 to 4 are identified as (1,1), (3,1), (1,3), and (5,1) modes [54]. As indicated in the figure, the frequencies of the first three modes in the ferrite slab are smaller than the FMR frequency ω_0. The mode having a frequency close to ω_0 and a sharp absorption dip, the (3,1) mode that is labeled 2 in Fig. 6.24, was selected for studies on ME coupling. The mode shows a 3 dB line width of 1 MHz that correspond to a quality factor $Q = 5 \times 10^4$.

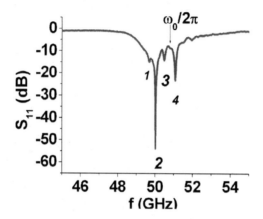

Figure 6.24 Magnetostatic forward volume modes in BaM [53]. See also Color Insert.

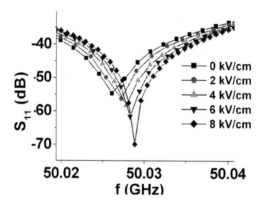

Figure 6.25 Frequency shift vs. E for mode 2 in Fig. 6.20 for BaM–PZT bilayer [55]. See also Color Insert.

The electric field effects on absorption profiles for the mode of interest are shown in Fig. 6.25. One notices an up-shift in the mode frequency with the application of E. Two significant changes are observed as E is increased from 0 to 8 kV/cm; a frequency shift of

4 MHz and an increase in absorption by 10–20 dB. The *E*-induced shift *δf* is much higher than the width of the excitation. Figure 6.26 shows similar data on the *E*-dependence of *δf* obtained for a bilayer with a ferrite thickness of 95 μm. Hysteresis and remanence are seen in the figure and the variation in *δf* with *E* tracks the dependence of the piezoelectric deformation on *E* in PZT. A maximum shift of 8 MHz is measured for *E* = ±10 kV/cm. Thus the ME coupling constant *A* = *δf*/*E* = 0.8 MHz cm/kV and must be compared with theoretical estimates of 22 MHz cm/kV. Possible causes of small shifts are discussed later. The main point here is mm-wave ME coefficients can be measured by FMR techniques.

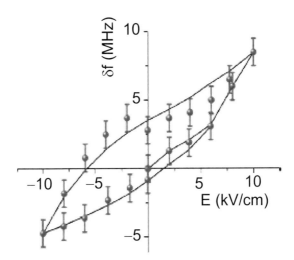

Figure 6.26 Shift vs. *E* for 95 μm thick BaM and PZT bilayer [55]. See also Color Insert.

6.6.8 Theory of MM-Wave ME Interactions

Preliminary studies are underway on mm-wave ME interactions in BaM–PZT. The primary objective is to determine the strength of ME interactions for the layered samples. The data in Figs. 6.24–6.26 constitute the first experiment on mm-wave ME effects in BaM–PZT. These effects are due to the influence of mechanical deformation on resonant frequencies of the modes in BaM via magnetostriction.

The effect can be estimated by considering the magnetostriction contribution to magnetic anisotropy energy [54]:

$$F_\sigma = -\frac{3}{2}\lambda\sigma\alpha_z^2 \tag{6.43}$$

where λ is magnetostriction constant, σ is tension (compression), and α_z is cosine of the angle between tension (compression) and direction of magnetization. According to the Hook's law, $\sigma = Y\Delta s/s$, where Y is the Young's modulus, $\Delta s/s$ is the tensile strain of the ferrite layer. The magnetostriction contribution to the anisotropy field can, therefore, be written as

$$\delta H_A = -\frac{1}{M}\frac{\partial F_\sigma}{\partial \alpha_z} = 3\frac{\lambda Y}{M}\frac{\Delta s}{s}\alpha_z \tag{6.44}$$

For barium ferrite $\lambda \approx -1.5 \times 10^{-5}$, M = 375 G, $Y \approx 1.3 \times 10^{12}$ dynes/cm^2, and maximum value $\Delta s/s \approx 5 \times 10^{-4}$ which is same as the value for PZT for $E = 10$kV/cm, $\alpha_z \approx 1$, one obtains a maximum value of $\delta H_A \approx 80$ Oe. This enhancement in H_A is substantial and corresponds to an ME coupling coefficient $A = 22$ MHz cm/kV. Although the measured A-value of 0.8 MHz cm/kV for BaM–PZT is quite small compared to the theoretical estimates, it compares favorably with measured values for YIG–PZT and YIG/PMN–PT [42]. One needs to have well bonded BaM–PZT interface to realize the predicted shift.

6.6.9 Theory of MM-Wave Hybrid Modes

Theory of mm-wave properties of hybrid ferromagnetic–ferroelectric layered structures is somewhat of a challenge. The main difficulties arise from the need to solve Maxwell equations in spatially inhomogeneous system where permeability μ of the ferrite layer and the permittivity ε of the ferroelectric are, tensors. An additional difficulty arises due to nonlinearities of μ and ε. Most cases cannot be solved exactly and one needs to develop approximate methods of investigation in this problem.

An approach that allows one to obtain analytical predictions on the microwave behavior of *unbound* hybrid structures was developed in [50]. In this work we obtained expressions for the hybrid spin-EM modes of YIG–BST, investigated the effect of E and H on the mode spectra, and determined the conditions for the best dual tunability of the modes. The approach is based on three tasks. First, we determined the static properties of the structure, such as equilibrium magnetization of YIG and static electric polarization of BST. Second, we found the resonance frequencies and distributions of electromagnetic field for individual modes of YIG and BST. Finally, using the coupled-mode theory, we derived expressions for the resonance frequencies of coupled modes. The resulting expressions have no adjustable parameters and demonstrate excellent agreement with our data [50]. But in our work in [50], we did not consider the anisotropy in YIG or mechanical contact between YIG and BST, and, therefore, did not take into account electrostrictive and magnetostrictive effects. This approximation is not valid in the case of highly anisotropic magnetic materials such as $Zn_2\mathbf{Y}$, Ba$\mathbf{M}$, or Sr$\mathbf{M}$ and one need to explicitly consider anisotropic properties of the ferrites.

6.7 CONCLUSIONS

There is a critical need for frequency tunable devices such as resonators, phase shifters, delay lines, and filters for next generation applications in the microwave and millimeter wave frequency regimes. These needs include conventional radar and signal processing devices as well as pulse based devices for digital radar and other systems applications. For secure systems, in particular, one must be able to switch rapidly between frequencies and to do so with a limited power budget. Traditional tuning methods with a magnetic field are slow and power consumptive. Electric field tuning offers new possibilities to solve both problems.

Ferrite–piezoelectric composites represent a promising new approach to build a new class of fast electric field tunable low power devices based on ME interactions. An electric field E applied to the piezoelectric transducer produces a mechanical deformation that couples to the ferrite and produces a shift in the resonance field. Unlike the situation when magnetic fields are used for such tuning, the process is fast because there are no inductors, and the power budget is small because the biasing voltages involve minimal currents. The critical goal for the future is in the development of a wide class of efficient wide band and low-loss electrically tunable magnetic film devices for battlefield radar, signal processing, and secure and experimental evaluation of characteristics. The anticipated advantages of ME devices are yet to be exploited.

Acknowledgments

Financial support for the research was provided by the Russian Foundation for Basic Research, and Programs of Russian Ministry of Education and Science.

References

1. "Magnetoelectric Interaction Phenomena in Crystals," Eds. A.J. Freeman, H. Schmid, Gordon and Breach, London, New Jersy, Paris, 1975, 228p.

2. G.A. Smolenskii, I.E. Chupis, "Magnetoelectrics," Adv. Phys. Sci., 137, 415 (1982) (in Russian).

3. "Magnetoelectric Substances," Eds. Yu.N. Venevtsev, V.N. Lyubimov, Nauka, Moscow, 1990, 184p (in Russian).

4. Proceedings of the 2nd International Conference on Magnetoelectnc Interaction Phenomena in Crystals (MEIPIC-2, Ascona), Eds. H. Schmid, A. Janner, H. Grimmer, J.-P. Rivera, Z.-G. Ye, Ferroelectrics, **161–162**, 1993, 748p.

5. Proceedings of the 3rd International Conference on Magnetoelectric Interaction Phenomena in Crystals (MEIPIC-3, Novgorod), Ed. M.I. Bichurin, Ferroelectrics, **204**, 1997, 356p.

6. Proceedings of the Fourth Conference on Magnetoelectric Interaction Phenomena in Crystals (MEIPIC-4, Veliky Novgorod), Ed. M.I. Bichurin, Ferroelectrics, **279–280**, 2002, 386p.

7. G. Harshe, J.P. Dougherty, R.E. Newnham, "Theoretical modelling of 3-0, 0-3 magnetoelectric composites," Int. J. Appl. Electromagn. Mater., **4**, 161 (1993).

8. M.I. Bichurin, V.M. Petrov, et al., "Magnetoelectric materials: technology features and application perspective," in Magnetoelectric Substances, Nauka, Russia, 118 (1990) (in Russian).

9. M.I. Bichurin, V.M. Petrov, I.A. Kornev, "Investigation of magnetoelectric interaction in composite," Ferroelectrics, **204**, 289 (1997).

10. M.I. Bichurin, V.M. Petrov, "Magnetic resonance in layered ferrite-ferrielectric structures," JETP, **58**, 2277 (1988).

11. M.I. Bichurin, O.S. Didkovskaya, V.M. Petrov, S.E. Sofronev, "Resonant magnetoelectric effect in composite materials," Izv. vuzov., ser. Physic., **1**, 121 (1985) (in Russian).

12. V. Gheevarughese, U. Laletsin, V. Petrov, G. Srinivasan, N. Fedotov, "Low-frequency and resonance magnetoelectric effects in lead zirconate titanate and single crystal nickel zinc ferrite bilayers," J. Mater. Res., **22**, 2130 (2007).

13. M.I. Bichurin, V.M. Petrov, "Influence of external electric field on magnetic resonance frequency in magnetic ferroelectrics," Ferroelectrics, **167**, 147 (1995).

14. V.M. Petrov, M.I. Bichurin. "Magnetic resonance in magnetoelectric composites," Abstracts of Int. Conf. "New materials and technology in electrical engineering," Lodz, Poland, **136** (1995).

15. M.I. Bichurin, I.A. Kornev, V.M. Petrov, I.V. Lisnevskaya, "Investigation of magneoelectric interaction in composite," Ferroelectrics, **204**, 269 (1997).

16. M.I. Bichurin, I.A. Kornev, V.M. Petrov, A.S. Tatarenko, Yu.V. Kiliba, G. Srinivasan "Theory of magnetoelectric effects at microwave frequencies in a piezoelectric/magnetostrictive multilayer composite," Phys. Rev., **B64**, 094409 (2001).

17. Yu. N. Kotyukov, "About single crystals ferrites magnetostriction constants measurement by ferromagnetic resonance method," Phys. Solid State, **8**, 1149 (1967) (in Russian).

18. S. Lopatin, I. Lopatina, I. Lisnevskaya, "Magnetoelectric PZT/ferrite composite materials," Ferroelectrics, **162**, 63−68 (1994).

19. A.G. Gurevich, "Magnetic Resonance in Ferrites and Antiferromagnets," Nauka, Moscow, 1973, 591p (in Russian).

20. M.I. Bichurin, V.M. Petrov, N.N. Fomich, "Magnetoelectric ferromagnetic susceptibility in microwave range and measuring methods," in Magnetoelectric Substances, Nauka, Russia, 1990, 67 (in Russian).

21. M.I. Bichurin, V.M. Petrov, "Composite magnetoelectrics: their microwave properties," Ferroelectrics, **162**, 33 (1994).

22. M.I. Bichurin, V.M. Petrov, Kiliba Yu.V., G. Srinivasan, "Magnetic and magnetoelectric susceptibilities of a ferroelectric. ferromagnetic composite at microwave frequencies," Phys. Rev., **B66**, 134404 (2002).

23. M.I. Bichurin, V.M. Petrov et al., "Resonance magnetoelectric effect in multilayer composites," Ferroelectrics, **280**, 187 (2002).

24. M.I. Bichurin, V.M. Petrov, D.A. Filippov, G. Srinivasan, "Influence of constant and ac electric fields on ferromagnetic resonance in magnetoelectric composites," Bull. Am. Phys. Soc., 223 (2004).

25. O.Yu. Belyaeva, L.K. Zarembo, C.H. Karpachev, "Magnetoacoustics ferrits and magnetoacoustics resonance," Adv. Phys. Sci., **162**, 107 (1992).

26. "Physical acoustics," Ed. W.P. Mason, v.III, Pt. B. Lattice dynamics, Academic Press, New York, 1965, 391p.

27. M.I. Bichurin, V.M. Petrov, I.A. Kornev, "Relaxation processes in magnetoelectric composites in magnetic resonance range," Bulletin of NovSU: ser. "Natural and technical sciences," 5, 3 (1997) (in Russian).

28. M.I. Bichurin, V.M. Petrov, G. Srinivasan, Yu.V. Kiliba, "Magnetic field sensors on the base of magnetoelectric composites," Abstracts of IV Int. conf. M., MIET, 2002. Part.1, 329.

29. M.I. Bichurin, V.M. Petrov, G. Srinivasan, Yu.V. Kiliba, "Microwave power sensors based on magnetoelectric composites," Abstracts of IV Int. conf. M., MIET, 2002. Part.1, 330.

30. K Yu. Viliba, M.I. Bichurin, V.M. Petrov, G. Srinivasan, J.V. Mantese, "Magnetoelectric composites for magnetic field and microwave power sensors," Bull. Am. Phys. Soc., 494 (2003).

31. M. Fiebig, "Revival of the magnetoelectric effect," J. Phys. D: Appl. Phys., **38**, R1 (2005).

32. Y.K. Fetisov, K.E. Kamentsev, A.Y. Ostashchenko, G. Srinivasan, "Wideband magnetoelectric characterization of a ferrite-piezoelectric

multilayer using a pulsed magnetic field," Solid State Comm., **132**, 13 (2004).

33. N.A. Spaldin, M. Fiebig, "The renaissance of magnetoelectric effect," Science **309**, 391 (2005).

34. N.A. Hill, "Why there are so few magnetic ferroelectrics," J. Phys. Chem., **B104**, 6694 (2000).

35. H. Schmid, "Introduction to complex mediums for optics and electromagnetics," Eds. W.S. Weiglhofer, A. Lakhtakia, SPIE Prsee, Bellingham, WA, 167 (2003).

36. V.E. Wood, A.E. Austin, "Possible applications for magnetoelectric materials," Proc. Symposium on Magnetoelectric Interaction Phenomena in Crystals, Seattle, May 21–24, 1973, eds. A.J. Freeman and H. Schmid, Gordon and Breach Science Publishers, New York, 181 (1975).

37. T.G. Lupeiko, I.V. Lisnevskaya, M.D. Chkheidze, B.I. Zvyagintsev, "Laminated magnetoelectric composites based on nickel ferrite and PZT materials," Inorganic Mater., **31**, 1139 (1995).

38. N. Cai, C.-W. Nan, J. Zhai, Y. Lin, "Large high-frequency magneto-electric response in laminated composites of piezoelectric ceramics, rare-earth iron alloys, and polymers," Appl. Phys. Lett., **84**, 35 (2004).

39. J. Li, I. Levine, J. Slutsker, V. Provenzano, P.K. Schenck, R. Ramesh, J. Ouyang, A.L. Roytburd, "Self-assembled multiferroic nanostructures in the CoFe2O4-PbTiO3 system," Appl. Phys. Lett., **87**, 072909 (2005).

40. G. Srinivasan, C.P. DeVreugd, V.M. Laletin, N.N. Paddubnaya, M.I. Bichurin, "Resonant magnetoelectric coupling in trilayers of ferromagnetic alloys and piezoelectric lead zirconate titanate: the influence of bias magnetic field," Phys. Rev., **B71**, 184423 (2005).

41. D.A. Filippov, M.I. Bichurin, V.M. Petrov, V.M. Laletin, N.N. Paddubnaya, G. Srinivasan, "Giant magnetoelectric effect in composite materials in the region of electromechanical resonance," Tech. Phys. Lett., **30**, 6 (2004).

42. S. Shastry, G. Srinivasan, M.I. Bichurin, V.M. Petrov, A.S. Tatarenko, "Microwave magnetoelectric effects in single crystal bilayers of

yttrium iron garnet and lead magnesium niobate-lead titanate," Phys. Rev., **B70**, 064416 (2004).

43. M.I. Bichurin, V.M. Petrov, O.V. Ryabkov, S.V. Averkin, G. Srinivasan, "Theory of magnetoelectric effects at magneto-acoustic resonance in single crystal ferromagnetic-ferroelectric heterostructures," Phys. Rev., **B72**, 060408 (R) (2005).

44. A.S. Tatarenko, M.I. Bichurin, G. Srinivasan, "Electrically tunable microwave filters based on ferromagnetic resonance in single crystal ferrite-ferroelectric bilayers," Elec. Lett., **41**, 596 (2005).

45. Y.K. Fetisov, G. Srinivasan, "Electrically tunable ferrite-ferroelectric microwave delay lines," Appl. Phys. Lett., **87**, 103502 (2005).

46. Y.K. Fetisov, G. Srinivasan, "A ferrite/piezoelectric microwave phase shifter: studies on electric field tunabulity," Elec. Lett., **41**, 1066 (2005).

47. Y.K. Fetisov, G. Srinivasan, "Electric field tuning characteristics of a ferrite-piezoelectric microwave resonator," Appl. Phys. Lett., **88**, 143503 (2006).

48. A.A. Semenov, S.F. Karmanenko, B.A. Kalinikos, G. Srinivasan, A.N. Slavin, J.V. Mantese, "Dual-tunable hybrid wave ferrite-ferroelectric microwave resonator," Elec. Lett., **42**, 641 (2006).

49. A.A. Semenov, S.F. Karmanenko, V.E. Demidov, B.A. Kalinikos, G. Srinivasan, A.N. Slavin, J.V. Mantese, "Ferrite-ferroelectric layered structures for electrically and magnetically tunable microwave resonators," Appl. Phys. Lett., **88**, 033530 (2006).

50. A.B. Ustinov, V.S. Tiberkevich, G. Srinivasan, A.N. Slavin, A.A. Semenov, S.F. Karmanenko, B.A. Kalinikos, J.V. Mantese, "Electric field tunable ferrite-ferroelectric hybrid wave microwave resonators: experiment and theory," J. Appl. Phys., **100**, 093905 (2006).

51. A.S. Tatarenko, G. Srinivasan, M.I. Bichurin, "A magnetoelectric microwave phase shifter," Appl. Phys. Lett., **88**, 183507 (2006).

52. G. Srinivasan, I.V. Zavislyak, A.S. Tatarenko, "Millimeter-wave magnetoelectric effects in bilayers of barium hexaferrite and lead zirconate titanate," Appl. Phys. Lett., **89**, 152508 (2006).

53. G. Srinivasan, I.V. Zavislyak, A.S. Tatarenko, "Millimeter-wave magnetoelectric effects in bilayers of barium hexaferrite and lead zirconate titanate," Appl. Phys. Lett., **89**, 152508 (2006).

54. I.V. Zavislyak, V.I. Kostenko, T.G. Chamor, L.V. Chevnyuk, "Ferromagnetic resonance in epitaxial films of uniaxial barium hexaferrites," Tech. Phys., **50**, 520 (2005).

55. A.M. Balbashov, L.N. Rybina, Y.K. Fetisov, V.F. Meshcheryakov, G. Srinivasan, "The floating zone crystal growth of Ni, Co, Ni-Co, Ni-Zn and Co-Zn ferrospinels under high oxygen pressure," J. Crys. Growth **275**, e733 (2005).

Chapter 7

Magnetoelectric Effects in Nanocomposites

V.M. Petrov and M.I. Bichurin

Institute of Electronic and Information Systems, Novgorod State University,
173003 Veliky Novgorod, Russia

This chapter dwells on distinctive features of magnetoelectric (ME) interactions in ferrite–PZT nanobilayers at low frequencies and in electromechanical resonance and ferromagnetic resonance regions. The models take into account the clamping effect of substrate, flexural deformations, and the contribution of lattice mismatch between composite phases and substrate to ME coupling. Lattice mismatch effect has been taken into account by using the classical Landau–Ginzburg–Devonshire phenomenological thermodynamic theory. The strength of low-frequency ME interactions is shown to be weaker than for thick film bilayers due to the strong clamping effects of the substrate by giving the example of NFO–PZT nanobilayer on $SrTiO_3$ substrates. However, flexural deformations result in the considerably lower rate of change of ME voltage coefficient with substrate thickness compared to the case when neglecting the flexural strains. To avoid the strong clamping effects of the substrate, nanopillars of a magnetostrictive material in a piezoelectric matrix can be used as an alternative. For nanopillars of NFO in PZT matrix on MgO, the substrate pinning effects are negligible when the length of the pillar is much greater than its radius.

Nanostructures in the shape of wires, pillars, and films are important for increased functionality in miniature devices [4]. A model of the static magnetoelectric (ME) effects in $BaTiO_3$–$CoFe_2O_4$ nanopillars and nanobilayers was considered in Chapter 4. However, a fundamental understanding of the size and shape-dependent characterization of the ME effect in composites, particularly down to nanoscale dimensions, is presently lacking. An improved knowledge of the nanoscale ME properties will help achieve miniaturization of potential ME devices.

In this chapter, we concentrate on distinctive features of ME effects in ME nanostructures. Important aspects of our approach are as follows. (1) Expressions for ME coefficients are obtained using the solution of elastostatic, electrostatic, and magnetostatic equations in terms of material parameters for bulk materials (piezoelectric coupling, magnetostriction, elastic constants, etc.). (2) Improved expressions for ME output take into account the clamping effect of substrate. (3) A further improvement of theoretical modeling the ME interaction with due regard for flexural deformations of the nanostructure. (4) Estimations for variation of material parameters due to lattice mismatch between composite phases and substrate. (5) A theoretical model for an enhancement of ME coupling at electromechanical resonance (EMR) is obtained using elastodynamic equations. (6) Reducing the operating EMR frequency is predicted for obtaining the giant ME effect by using bending modes.

7.1 LOW-FREQUENCY ME EFFECT IN NANOBILAYER ON SUBSTRATE

This part is focused on modeling the ME effects in a ferrite–piezoelectric nanobilayer on a substrate. We have chosen nickel ferrite (NFO)–lead zirconate titanate (PZT) as the model system. Although NFO has a weaker magnetostriction as compared to cobalt ferrite, it has superior piezomagnetic coupling which is a key ingredient for strong ac ME interactions. Similarly, PZT has much stronger piezoelectric effects than barium titanate. We consider nanostructures grown on MgO substrates [8, 9]. MgO is experimentally attractive because of its chemical inertness and good lattice match with the ferroelectric and ferromagnetic oxides, particularly the latter. The ME voltage coefficients α_E have been estimated for field orientations corresponding to minimum demagnetizing fields and maximum α_E. The effect of substrate or template clamping has been described in terms of dependence of α_E on substrate/template dimensions or volume fraction. In the case of bilayers, α_E drops exponentially with increasing volume of MgO.

For the estimation of α_E it is assumed that the piezoelectric phase is electrically poled in a dc field E and that the composite is subjected to a bias field H and an ac magnetic field δH, giving rise to a piezomagnetic deformation. We then solve magnetostatic and elastostatic equations in NFO, and elastostatic and electrostatic equations in PZT, taking into account boundary conditions. Then the ME voltage coefficient which is the ratio of ME susceptibility and permittivity is calculated. The model can be used to estimate the ME couplings from known material parameters (piezoelectric coefficients, magnetostriction, stiffness, etc.) or data on ME coupling can be used to extract composite parameters.

For a polarized piezoelectric phase with the symmetry ∞m and magnetostrictive phase with cubic symmetry, the strains and electric displacement are determined by Eqs. 2.7–2.9. The law of elasticity for the substrate has the form:

$$^sS_i = {}^ss_{ij}\,{}^sT_j, \tag{7.1}$$

where ^sS_i and ^sT_j are strain and stress tensor components, and s_{ij}, compliance coefficients of the substrate, respectively.

In Chapter 2, it was shown that α_E is expected to be maximum for in-plane longitudinal fields, i.e., the dc bias magnetic field and the ac electric and magnetic field, are all parallel to each other and to the sample plane. For the two other field orientations, out-of-plane longitudinal and in-plane transverse fields, α_E is predicted to be weak due to combination of weak piezomagnetic and piezoelectric couplings and demagnetizing fields. These predictions are in general agreement with data. Thus the estimates of α_E here for a NFO–PZT bilayer on MgO is limited to in-plane longitudinal fields. We assume a coordinate system with the sample in the (1,2) plane, and that the piezoelectric phase is poled with an electric field E along the direction-1 and that a bias magnetic field H_1 and an ac field δH_1 are applied along the same direction-1. The resulting ac electric field is measured along the same direction for the determination of the ME voltage coefficient $\alpha_{E,11} = \delta E_1/\delta H_1$. For finding $\alpha_{E,11}$, the following boundary conditions are used:

$$^pS_1 = {}^mS_1 = {}^sS_1,$$
$$^pS_2 = {}^mS_2 = {}^sS_2,$$
$$^pT_1 \upsilon + {}^mT_1 (1 - \upsilon) + {}^sT_1 \upsilon_s = 0, \qquad (7.2)$$
$$^pT_2 \upsilon + {}^mT_2 (1 - \upsilon) + {}^sT_2 \upsilon_s = 0,$$

where $\upsilon = {}^p\upsilon/({}^p\upsilon + {}^m\upsilon)$, $\upsilon_s = {}^s\upsilon/({}^p\upsilon + {}^m\upsilon)$ and $^p\upsilon$, $^m\upsilon$, and $^s\upsilon$ denote the volume of piezoelectric and magnetostrictive phases and substrate, respectively.

The ME voltage coefficient is calculated numerically from Eq. 7.2 taking into account Eqs. 2.7–2.9 and Eq. 7.1 and using the open circuit condition $D_1 = 0$. Figure 7.1 shows the PZT volume fraction dependence of ME voltage coefficient for NFO–PZT on MgO that is characterized by material parameters listed in Table 2.4. For a free-standing nanobilayer, $\alpha_{E,11}$ increases with increasing PZT volume, attains a peak value of 1.6 V/cmOe for $\upsilon = 0.11$, and then drops rapidly with increasing υ. A dramatic, factor-of-two decrease in $\alpha_{E,11}$ is expected when the film is deposited on a thin MgO substrate of volume that is only 50% of the film volume. As seen in Fig. 7.1, further increase in υ_s leads to a substantial decrease in $\alpha_{E,11}$ and the ME coupling vanishes when the film is assumed to be on MgO of volume fraction 10 or more.

Figure 7.2 shows the variation in the peak value of $\alpha_{E,11}$ with the substrate volume fraction υ_s along with the PZT volume fraction υ corresponding to peak-$\alpha_{E,11}$. The up-shift of the PZT volume fraction that results in maximal ME effect is stipulated by variation of effective compliance of the sample with the substrate thickness.

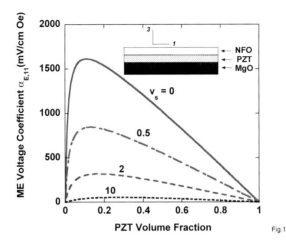

Figure 7.1 Diagram showing a nickel ferrite (NFO)–lead zirconate titanate (PZT) nanobilayer in the (1,2) plane on MgO substrate. It is assumed that PZT is poled with an electric field E_1 along 1, the bias magnetic field H_1 and ac magnetic field δH_1 are along axis-1, and the ac electric field δE_1 is measured along direction-1. Estimated PZT volume fraction dependence of in-plane longitudinal ME voltage coefficient $\alpha_{E,11} = \delta E_1/\delta H_1$ is shown for a series of volume fraction v_s for MgO. See also Color Insert.

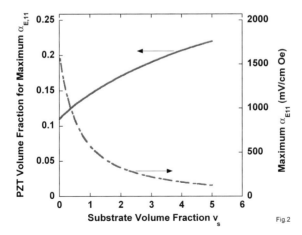

Figure 7.2 Variation with substrate volume fraction v_s of the peak value of ME voltage coefficient $\alpha_{E,11}$ in Fig. 7.3 and the corresponding PZT volume fraction. See also Color Insert.

7.2 FLEXURAL DEFORMATION OF ME NANOBILAYER ON SUBSTRATE

Theoretical modeling of low-frequency ME effect described above is based on the homogeneous longitudinal strain approach. However, configurational asymmetry of a bilayer implies bending the sample in applied magnetic or electric field and variation in ME response. One of the principal objective of present section is modeling of the ME interaction in a magnetostrictive–piezoelectric nanobilayer on a substrate taking into account the flexural strains [3]. NFO–PZT on $SrTiO_3$ substrate is chosen as the model system for numerical estimations. The nanobilayers of NFO and PZT and was recently epitaxially grown on $SrTiO_3$ substrate and their ME coupling was experimentally observed [5]. We calculated ME voltage coefficients α_E for transverse field orientations to provide minimum demagnetizing fields and maximum α_E (see Fig. 7.4).

We assume the symmetry of piezoelectric to be ∞m and that of piezomagnetic to be cubic. As shown in Fig. 7.3, x_m, x_p, and x_s are the neutral axes which are located at the horizontal mid-plane of piezomagnetic, piezoelectric, and substrate layers, respectively, and separated by a distances h_m and h_p.

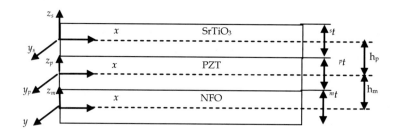

Figure 7.3 Schematic of a bilayered structure on a substrate.

The thickness of the plate is assumed small compared to remaining dimensions. We assume the longitudinal axial strains of each layer to be linear functions of the vertical coordinate z_i to take into account bending the sample. To preserve force equilibrium, the axial forces in the three layers add up to zero, that is,

$$F_{p1} + F_{m1} + F_{s1} = 0, \tag{7.3}$$

where $F_{i1} = \int_{-{}^{i}t/2}^{{}^{i}t/2} {}^{i}T_1 dz_1$.

The moment equilibrium condition has the form:

$$F_{m1}h_m + F_{p1}(h_m + h_p) = M_{m1} + M_{p1} + M_{s1}, \tag{7.4}$$

where $M_{i1} = \int_{-{}^{i}t/2}^{{}^{i}t/2} z_i {}^{i}T_1 dz_1$.

Simultaneous solving Eqs. 7.3 and 7.4 taking into account Eqs. 2.7–2.9 and 7.1 enables finding the axial stress components in the piezoelectric layer ${}^{p}T_1$ and ${}^{p}T_2$. Then the expression for ME voltage coefficient can be expressed using the open circuit condition

$$\alpha_{E31} = \frac{E_3}{H_1} = -\frac{{}^{p}d_{31} \int_{-{}^{p}t/2}^{{}^{p}t/2} ({}^{p}T_1 + {}^{p}T_2)dz}{t\, H_1\, {}^{p}\varepsilon_{33}}, \tag{7.5}$$

where $t = {}^{m}t + {}^{p}t + {}^{s}t$ is the total thickness of considered structure.

In case of neglecting the flexural strains, it is easily shown that expression for ME voltage coefficient reduces to well-known expression of Part 2 which was obtained with the assumption of homogeneous longitudinal strains.

7.3 LATTICE MISMATCH EFFECT

It is well known that the lattice mismatch between the substrate and piezoelectric layers results in variation of piezoelectric coefficients and permittivity. This variation can be found using the Landau–Ginzburg–Devonshire phenomenological thermodynamic theory [1, 7]. According to this approach, the thermodynamic potential G' of a thin film on a thick substrate is defined as

$$G' = G + S_{s1}T_{s1} + S_{s2}T_{s2} + S_{s6}T_{s6}, \tag{7.6}$$

where G is the elastic Gibbs function for the PZT layer without substrate, S_{s1}, S_{s2}, and S_{s6} are in-plane strains at the film/substrate interface arising from lattice mismatch, T_{s1}, T_{s2}, and T_{s6} are residual stress components. In case of (001) ferroelectric thin film grown in a cubic paraelectric phase on a cubic (001) substrate, the expression for G is as follows:

$$G = a_1 P_3^2 + a_{11} P_3^4 + a_{111} P_3^6 - \frac{1}{2} s_{11}(T_{s1}^2 + T_{s2}^2) - s_{12} T_{s1} T_{s2}$$
$$- Q_{12}(T_{s1} + T_{s2})P_3^2, \tag{7.7}$$

where a_1, a_{11}, and a_{111} are ferroelectric dielectric stiffnesses, Q_{12} is electrostrictive coupling coefficient.

The stress components can be found by using the mechanical conditions: $\partial G/\partial_{Ts1} = -S_{s1}$ and $\partial G/\partial T_{s2} = -S_{s2}$. Taking into consideration that $S_6 = 0$ and $S_1 = S_2 = S_m$, the misfit strain $S_m = (b - a_0)/b$ can be calculated using the substrate lattice parameter b and the equivalent cubic cell constant a_0 of the free-standing film. For the PZT film on $SrTiO_3$ substrate $a_0 = 0.397$ nm and $b = 0.393$ nm. Using these equations with parameters [2] of the Gibbs function enables determining the dielectric constant and piezoelectric coefficients of PZT: ${}^P\varepsilon_{11}/\varepsilon_0 = 51$, ${}^P d_{31} = -18$ pm/V. Similarly, the piezomagnetic parameters of the NFO film should be less compared to bulk NFO. Nevertheless, in the NFO/PZT/STO composite film, the bottom PZT layer acts as a buffer layer and effectively reduces constrains from the STO substrate and compressive strains in the NFO layer are almost released [2]. In what follows, we neglect the contribution of residual strains on NFO parameters arising from lattice mismatch.

Substituting the appropriate material property values and the dimensions of the structure into Eq. 7.5, we can find the ME voltage coefficient with regard to axial and bending stresses for the bilayer on a substrate.

Our model predicts decreasing the ME voltage coefficient for the free-standing bilayer compared to the case when neglecting the flexural strains that is connected with bending the sample and decreasing the strains of piezoelectric layer. Placing the bilayer on a substrate gives rise to an increase of ME output with thickness of bilayer due to sign reversal in the contribution of flexural strain to the total strain of PZT in comparison with the free-standing bilayer. Further increase in v_s leads to a substantial decrease in ME coupling as shown in Fig. 7.4. However, rate of change of $\alpha_{E,31}$ is considerably lower than that for the case when neglecting the flexural strains. It can be accounted for by less force which is required for flexural straining the thick substrate compared to longitudinal one.

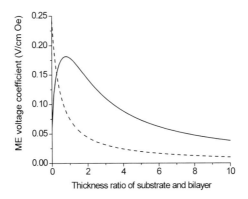

Figure 7.4 Dependence of ME voltage coefficient $\alpha_{E,31}$ for NFO–PZT on SrTiO$_3$ substrate on ratio of substrate thickness to bilayer thickness for $v = 0.6$ calculated using Eq. 7.12 (solid line) and using the model of axial strains (dash line).

It should be noted that the experimentally measured ME effect in nanobilayer of NFO and BaTiO$_3$ (BTO) on SrTiO$_3$ substrate [2] exceeds the estimated one. The details about the nature of the ME coupling in the epitaxial BTO–NFO composite films are currently unknown and remain in need of investigation.

7.4 ME EFFECT IN A NANOPILLAR

It is clear from the results in Figs. 7.1 and 7.4 that ME properties of mechanically coupled bilayer will be negatively affected by substrate clamping. Therefore, as an alternative, we have modeled magnetostrictive nanopillars grown in a piezoelectric matrix. Such heterostructures can still be limited by clamping due to the substrate on which they are grown. Nanopillars of CoFe$_2$O$_4$ in BaTiO$_3$ matrix on (001) SrTiO$_3$ substrates was demonstrated by Zheng *et al.* [4]. We consider here a similar nanopillar of NFO and PZT. The unit of this structure can be modeled by two coaxial cylinders consisting of ferrite and PZT on MgO, as in Fig. 7.5. In terms of mechanical connectivity such a structure is also known as a 3–1 composite. Directions of polarization and ac electric and ac and dc magnetic fields are supposed to coincide with the axis of the cylinders. Taking into account the axial symmetry, the elastostatic equation for this case is written in cylindrical coordinates.

$$\frac{\partial T_{rr}}{\partial r} + \frac{1}{r}(T_{rr} - T_{\theta\theta}) = 0. \tag{7.8}$$

Transferring Eqs. 2.7–2.9 and 7.1 to cylindrical coordinates by means of tensor relations, one finds for the stress tensor component in terms of strain tensor component. Substituting these expressions into Eq. 7.8 enables obtaining the following equation for the radial displacement u_r that defines strain components $S_{rr} = \partial u_r/\partial r$ and $S_{\theta\theta} = u_r/r$ for both phases and MgO substrate:

$$\frac{\partial^2 u_r}{\partial r^2} + \frac{1}{r}\frac{\partial u_r}{\partial r} - \frac{u_r}{r^2} = 0. \tag{7.9}$$

Solving Eq. 7.9 and using the boundary conditions for displacement and stress components one can get numerical solution for the ME voltage coefficient $\alpha_{E,33}$ [7]. The computation results are shown in Fig. 7.5 for NFO–PZT pillars on MgO.

Results in Fig. 7.5 indicate no clamping due to MgO when the length of the pillar L_m is greater than its radius R. The magnitude of $\alpha_{E,33}$ and its variation with PZT volume fraction v is similar to that of in-plane ME interactions in a free-standing bilayer. For the nanopillar, substrate clamping and demagnetization effects become important only when $L_m < R$ and substrate thickness $L_s >> L_m$, leading to a substantial reduction in $\alpha_{E,33}$.

It is appropriate here to compare the results in Fig. 7.5 with the model discussed in Section 4.1 for static ME effects in CoFe$_2$O$_4$–BaTiO$_3$ nanopillars that was based on the Green's function approach. The model compared the estimates on ME coupling in bulk composites, bilayers, and nanopillars of the ferrite in barium titanate matrix. It accounted for the experimentally observed magnetic field induced polarization that was attributed to enhancement in the elastic coupling. Our results in Fig. 7.5 also predict strong ME coupling in nanopillars that will not be influenced by the substrate coupling when the pillar height is much greater than its radius, in agreement with results in Section 4.1.

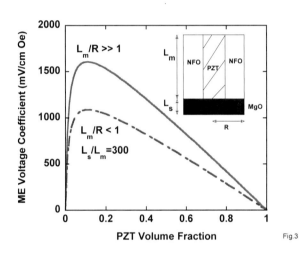

Figure 7.5 A nanopillar of NFO and PZT of radius R and length L_m on an MgO substrate of thickness L_s. All the electric and magnetic fields are assumed to be along the axis (direction-3) of the pillars. Estimated PZT volume fraction dependences of ME voltage coefficient $\alpha_{E,33}$ for NFO–PZT nanopillar are shown. See also Color Insert.

7.5 TRANSVERSE ME EFFECT AT LONGITUDINAL MODE OF EMR IN NANOBILAYER ON SUBSTRATE

This section is focused on modeling of the ME effect in ferrite–piezoelectric nanobilayers in EMR region. As an example, we obtained numerical estimations for the bilayer of CFO and barium titanate grown on $SrTiO_3$ substrate. The ME voltage coefficients α_E have been estimated for transverse field orientations corresponding to minimum demagnetizing fields and maximum α_E. As a model, we consider a ferrite–piezoelectric layered nanostructure on a substrate in the form of a thin plate with the length L. For the estimation of α_E it is assumed that the composite is subjected to a bias field H_0 and an ac magnetic field H, giving rise to a piezomagnetic deformation. Due to magnetostriction, the alternating magnetic field induces oscillations that propagate both in depth and in plane of the plate. In what follows, we consider oscillations of the lowest frequency, i.e., the volume oscillations propagating along the boundary.

We solve the equation of medium motion taking into account the magnetostatic and elastostatic equations, constitutive equations, Hooke's law, lattice mismatch effect, and boundary conditions. The equation of medium motion has the form:

$$\frac{\partial^2 u_1}{\partial x^2} = -k^2 u_1,$$

(7.10)

where u_1 is displacement in the traveling direction x. The wave value k is defined by expression:

$$k = \omega \sqrt{\left[{}^p\rho\, v + {}^m\rho(1-v) + {}^s\rho\, {}^s v\right] \left[\frac{v}{{}^p S_{11}} + \frac{1-v}{{}^m S_{11}} + \frac{v_s}{{}^s S_{11}}\right]^{-1}},$$

(7.11)

where ω is the circular frequency, ${}^p\rho$, ${}^m\rho$, and ${}^s\rho$ are the piezoelectric, piezomagnetic, and substrate densities, where $v = {}^p v/({}^p v + {}^m v)$, $v_s = {}^s v/({}^p v + {}^m v)$ and ${}^p v$, ${}^m v$, and ${}^s v$ denote the volume of piezoelectric and magnetostrictive phases and substrate, respectively.

For the solution of the Eq. 7.10, the following boundary conditions are used:

$${}^p S_1 = {}^m S_1 = {}^s S_1,$$
$${}^p T_1 v + {}^m T_1 (1-v) + {}^s T_1 v_s = 0.$$

(7.12)

The ME voltage coefficient $\alpha_{E\,13} = E_3/H_1$ is calculated numerically from Eqs. 2.7–2.9, 7.1, 7.10, and 7.12 and using the open circuit condition $D_3 = 0$. To take into consideration the energy loss, we set ω equal to $\omega' - i\omega''$ with $\omega''/\omega' = 10^{-3}$.

The resonance enhancement of ME voltage coefficient for free-standing bilayer is obtained at antiresonance frequency. For a free-standing nanobilayer, $\alpha_{E\,13}$ increases with increasing barium titanate volume, attains a peak value of 280 V/cmOe for $\upsilon = 0.5$, and then drops with increasing υ. A dramatic decrease in $\alpha_{E\,13}$ is shown in Fig. 7.6 when the CFO–BaTiO$_3$ film is deposited on a SrTiO$_3$ substrate. This decrease is stipulated by (i) clamping effect caused by the substrate and (ii) variation in piezoelectric and piezomagnetic coefficients and permittivity of components due to lattice mismatch.

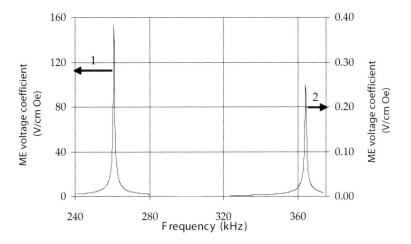

Figure 7.6 Frequency dependence of ME voltage coefficient $\alpha_{E,13} = \delta E_3 / \delta H_1$ for the free-standing bilayer (1) and for the bilayer on SrTiO$_3$ substrate with $\upsilon_s = 5$ (2). Calculations are for $\upsilon = 0.5$.

Figure 7.7 shows the variation in the peak value of $a_{E\,13}$ with the substrate volume fraction υ_s along with EMR frequency corresponding to peak $a_{E\,13}$. As seen in Fig. 7.9, further increase in υ_s leads to a substantial decrease in $a_{E\,13}$ and the ME coupling vanishes when the film is assumed to be on SrTiO$_3$ of volume fraction 20 or more. The up-shift of the EMR frequency that results in maximal ME effect is stipulated by variation of effective compliance of the sample with the substrate thickness.

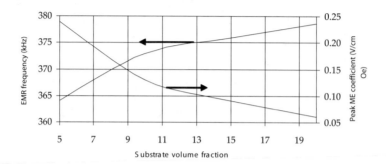

Figure 7.7 Variation of the peak value of ME voltage coefficient $a_{E,13}$ and the corresponding EMR frequency with substrate volume fraction v_s.

Thus, a model has been developed for ME interactions in a ferrite–piezoelectric nanobilayer on a substrate. For the CFO–BaTiO$_3$ bilayer on the substrate with increasing thickness, (i) the ME coefficient drops exponentially and (ii) EMR frequency increases.

7.6 TRANSVERSE ME EFFECT AT BENDING MODE OF EMR IN NANOBILAYER ON SUBSTRATE

A key drawback of resonance ME effect at longitudinal modes of EMR is that the resonance frequencies are quite high, on the order of hundreds of kHz, for nominal sample dimensions. At such frequencies, the eddy current losses for the magnetostrictive phase can be quite high, in particular for earth rare alloys such as Terfenol-D, resulting in an inefficient ME energy conversion. To reduce the operating frequency, one can use a strong ME coupling at bending modes of EMR [3].

This section presents a theoretical model of small-amplitude flexural oscillations of a bilayer plate, which is formed by magnetostrictive and piezoelectric nanolayers grown on a substrate. An in-plane bias field is assumed to be applied to magnetostrictive component to avoid the demagnetizing field. The thickness of the plate is assumed to be small compared to remaining dimensions. Moreover, the plate width is assumed small compared to its length.

In that case, we can consider only one component of strain and stress tensors in the EMR region. The equation of bending motion of bilayer has the form:

$$\nabla^2\nabla^2 w + \frac{\rho\, b}{D}\frac{\partial^2 w}{\partial \tau^2} = 0,$$

(7.13)

where $\nabla^2\nabla^2$ is biharmonic operator, w is the deflection (displacement in z-direction), t and ρ are thickness and average density of sample, $b = {}^Pt + {}^mt + {}^st$, $\rho = ({}^P\rho\,{}^Pt + {}^m\rho\,{}^mt + {}^s\rho\,{}^st)/b$, ${}^P\rho$, ${}^m\rho$, ${}^s\rho$, and ${}^Pt$, ${}^mt$, st are densities and thicknesses of piezoelectric and piezomagnetic layers and substrate, correspondingly, and D is cylindrical stiffness.

The boundary conditions for $x = 0$ and $x = L$ have to be used for finding the solution of above equation. Here L is length of bilayer. As an example, we consider the plate with free ends. At free end, the turning moment M_1 and transverse force V_1 equal zero:

$$M_1 = 0 \text{ and } V_1 = 0 \text{ at } x = 0 \text{ and } x = L,$$

(7.14)

where $M_1 = \int_A z T_1 dz_1$, $V_1 = \dfrac{\partial M_1}{\partial x}$, and A is the cross-sectional area of the sample normal to the x-axis.

We are interested in the dynamic ME effect; for an ac magnetic field H applied to a biased sample, one measures the average induced electric field E and calculates the ME voltage coefficient $a_E = E/H$. Using the open circuit condition, the ME voltage coefficient can be found as

$$\alpha_{E\,31} = \frac{E_3}{H_1} = -\frac{\displaystyle\int_{z_0-{}^Pt}^{z_0} {}^P E_3 dz}{t\, H_1},$$

(7.15)

where E_3 and H_1 are the average electric field induced across the sample and applied magnetic field. The energy losses are taken into account by substituting ω for complex frequency $\omega' + i\omega''$ with $\omega''/\omega' = 10^{-3}$.

As an example, we apply Eq. 7.15 to the bilayer of NFO and PZT. Figure 7.8 shows the frequency dependence of ME voltage coefficient at bending mode for free-standing bilayer. Graph of $\alpha_{E,31}$ reveals a giant value $\alpha_{E\,31}$ = 6.6 V/cmOe and resonance peak lies in the infralow frequency range.

Placing the bilayer of NFO and PZT on a $SrTiO_3$ substrate results in variations in ME coupling as in Fig. 7.9. At small values of substrate thickness, ME output considerably reduces and becomes zero at thickness ratio of substrate and bilayer V_s = 0.3. This is accounted for by decreasing the average stress across the piezoelectric component. The fact is that one part of this layer is subject to tension stress, while the other part suffers the compressive one. According to Eq. 7.15, the induced voltage is determined by average stress.

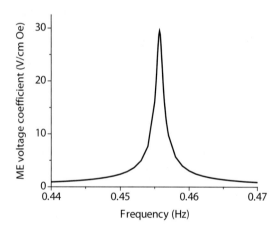

Figure 7.8 Theoretical frequency dependence of transverse ME voltage coefficients for a NFO–PZT bilayer for v = 0.67.

It should be noted that peak ME voltage coefficient at bending mode exceeds that at longitudinal mode in more than two times for V_s > 2 as in Fig. 7.9. The EMR frequency near-linearly depends on substrate thickness.

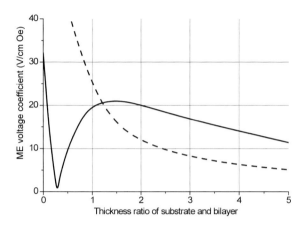

Figure 7.9 Dependence of peak ME voltage coefficient for the bilayer of NFO and PZT at bending mode on ratio of $SrTiO_3$ substrate thickness to bilayer thickness are presented for PZT volume fraction 0.6. For comparison, the ME voltage coefficient at longitudinal mode is shown (dash line). See also Color Insert.

7.7 ME EFFECT IN FERRITE–PIEZOELECTRIC NANOBILAYER AT FERROMAGNETIC RESONANCE

Investigations of ferromagnetic resonance (FMR) line shift by an applied electric field are described in Chapter 6 for layered ferrite–piezoelectric structures. In this chapter, we present a detailed treatment for electric field induced resonance field shift for FMR in layered structures taking into account the contribution of flexural deformations and effect of substrate clamping.

The layers are assumed to be perfectly bonded together; the ferrite and the substrate therefore restrict the deformation of the piezoelectric layer in an external electric field. The shearing forces of each layer produce bending moments since these forces are not applied centrally [3]. The flexural deformations can be taken into account by assuming the longitudinal axial strains of each layer to be linear functions of the vertical coordinate z_i:

$$^iS_j = {}^mS_{j0} + z_m/R_j,$$ (7.16)

where $^iS_{10}$ and $^iS_{20}$ are the centroidal strains along x- and y-axis at $z_i = 0$ and R_1 and R_2 are the radiuses of curvature, z_i is measured relative to i-layer ($i = m, p, s$).

One can conclude from geometric consideration that $^mS_{j0} - {}^pS_{j0} = h_m/R_j$ and $^pS_{j0} - {}^sS_{j0} = h_p/R_j$ where $h_m = ({}^mt + {}^pt)/2$, $h_p = ({}^pt + {}^st)/2$, ^mt, ^pt, and st are the thicknesses of piezomagnetic, piezoelectric, and substrate layers.

According to equilibrium conditions, the sum of axial forces in the three layers must be zero and the sum of the moments must equal the sum of moment-couples created by the shearing forces acting centrally on each layer:

$$F_{m1} + F_{p1} + F_{s1} = 0,$$
$$F_{m2} + F_{p2} + F_{s2} = 0,$$
$$F_{m1}h_m + F_{p1}(h_m + h_p) = M_{m1} + M_{p1} + M_{s1},$$ (7.17)
$$F_{m2}h_m + F_{p2}(h_m + h_p) = M_{m2} + M_{p2} + M_{s2},$$

where $\quad F_{i1} = \int_{-{}^it/2}^{{}^it/2} {}^iT_1 dz_i, \quad F_{i2} = \int_{-{}^it/2}^{{}^it/2} {}^iT_2 dz_i, \quad M_{i1} = \int_{-{}^it/2}^{{}^it/2} z_i {}^iT_1 dz_i,$ and

$$M_{i2} = \int_{-{}^it/2}^{{}^it/2} z_i {}^iT_2 dz_i.$$

Solving Eqs. 7.16 and 7.17 enables finding ^iT_j. Then Eqs. 7.16 can be solved for $^mS_{10}$, $^mS_{20}$, R_1, and R_2. Using the obtained centroidal strains and radiuses of curvature allows determining the axial stresses $^mT_1$ and $^mT_2$ from Eqs. 7.11. Next, one can substitute $^mT_1$ and $^mT_2$ into Eq. 6.16 and obtain the shift of magnetic resonance line taking into consideration Eq. 6.13.

We consider now a specific case of magnetic field H along <111>. For this case, matrix β has the form:

$$\beta_{ij'} = \begin{bmatrix} \frac{\sqrt{2}}{2} & \frac{\sqrt{6}}{6} & \frac{\sqrt{3}}{3} \\ -\frac{\sqrt{2}}{2} & \frac{\sqrt{6}}{6} & \frac{\sqrt{3}}{3} \\ 0 & -\frac{\sqrt{6}}{3} & \frac{\sqrt{3}}{3} \end{bmatrix}$$ (7.18)

and the geometrical demagnetization factors are $N_{11}^F = N_{22}^F = 0$, $N_{33}^F = 4\pi$.

Now we apply the model to the ferrite–piezoelectric bilayer of PMN–PT and yttrium iron garnet (YIG). The shift of magnetic resonance line was estimated at 9.3 GHz for the following parameters:

$$PMN\text{–}PT: {}^p d_{310} = -600 \cdot 10^{-12} \text{ m/V}, {}^p s_{11} = 23 \cdot 10^{-12} \text{ m}^2/\text{N};$$
$${}^p s_{12} = -8.3 \cdot 10^{-12} \text{ m}^2/\text{N}; \varepsilon_{330}/\varepsilon_0 = 4100.$$

$$YIG: {}^m t = 4.7 \; \mu\text{m}; \lambda_{100} = -1.4 \cdot 10^{-6}; \lambda_{111} = -2.4 \cdot 10^{-6};$$
$$4\pi M_0 = 1750 \text{ G};$$
$$H_a = -42 \text{ Oe}; {}^m s_{11} = 4.8 \cdot 10^{-12} \text{ m}^2/\text{N}; {}^m s_{12} = -1.4 \cdot 10^{-12} \text{ m}^2/\text{N}.$$

Figure 7.10 shows the calculated YIG volume fraction dependence of shift of magnetic resonance line for free-standing bilayer of YIG and PMN–PT. The similar dependence was obtained by using our previous model that does not take into account the flexural deformations. This dependence is given in Fig. 7.10 for comparison. It should be noted that the proposed model reduces to our previous model by letting the radiuses of curvature in Eq. 7.16 go to infinity.

It is evident that the strength of ME coupling is determined by the stress induced in ferrite layer. According to our previous model, a strong microwave ME interaction is expected when the volume fraction of the piezoelectric phase is sufficiently high, the piezoelectric component has a large piezoelectric coupling coefficient, and the magnetic phase has a small saturation magnetization and high magnetostriction [6]. In addition, the present studies indicate that flexural deformation gives rise to a decrease in the stress of ferrite layer and a decrease in the strength of ME interaction for a free-standing bilayer as in Fig. 7.10. The proposed model and our previous model result the nearly equal values for $\upsilon \ll 1$ and $(1 - \upsilon) \ll 1$ (where υ is YIG volume fraction).

We have a completely different situation if the bilayer is placed on a substrate [10]. The obtained shift of magnetic resonance line increases for $\upsilon_s > 0.5$ as in Fig. 7.11, where υ_s is the substrate thickness to bilayer thickness ratio.

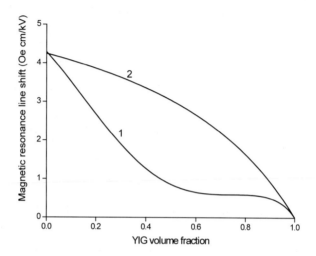

Figure 7.10 YIG volume fraction dependence of magnetic resonance line shift at E = 1 kV/cm for free-standing bilayer of YIG and PMN–PT taking into account (1) and ignoring (2) flexural deformations.

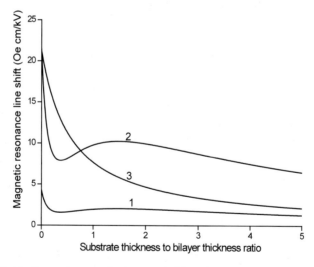

Figure 7.11 Dependence of electric field induced shift of magnetic resonance line on substrate thickness to bilayer thickness ratio for YIG/PMN–PT bilayer on GGG substrate taking into account flexural deformations for E = 1 kV/cm (1) and E = 5 kV/cm (2) and ignoring flexural deformations (3).

The bilayer on a substrate with thickness that is approximately equal to that of the bilayer reveals an increase in the ME effect by a factor of two. It is caused by the sign reversal in the contribution of flexural strain. The stresses in the bilayer and the substrate are redistributed so that Z-dependent axial stress of ferrite is of the same sign as the substrate and the average stress of ferrite becomes maximum for a substrate thickness equal to that of the bilayer. Further increase in the substrate thickness leads to a substantial decrease in electric field induced stress of ferrite film due to clamping effect produced by the substrate. However, rate of decrease in ME coupling strength is considerably lower than that for the case when neglecting the flexural strains. It can be accounted for by less force which is required for flexural straining the thick substrate compared to longitudinal deformation.

7.8 CONCLUSIONS

The models have been developed describing the distinctive features of ME interactions in ferrite–PZT nanobilayers at low frequencies and in EMR and FMR regions. Expressions for ME coefficients are obtained using the solution of elastostatic/elastodynamic and electrostatic and magnetostatic equations taking into account the clamping effect of substrate, flexural deformations, and the contribution of lattice mismatch between composite phases and substrate to ME coupling. The ME voltage coefficients are estimated from known material parameters.

For a NFO–PZT nanobilayer on $SrTiO_3$ substrates, the strength of low-frequency ME interactions is shown to be weaker than for thick film bilayers due to the strong clamping effects of the substrate. However, flexural deformations result in the considerably lower rate of change of ME voltage coefficient with substrate thickness compared to the case when neglecting the flexural strains. For increasing the ME coupling, the nanobilayer can be constrained by a clamper which produces the z-component of stress in the composite (perpendicular to the interface). In this case, the substrate clamping considerably reduces and ME effect is determined by longitudinal piezoelectric and piezomagnetic coefficients. However, the described structure seems to be too complex and hardly realizable with regard to nanobilayers.

The peak ME voltage coefficient at bending mode exceeds that at longitudinal mode in more than two times for thick enough substrate.

To avoid the strong clamping effects of the substrate, nanopillars of a magnetostrictive material in a piezoelectric matrix can be used as an alternative. For nanopillars of NFO in PZT matrix on MgO, the substrate pinning effects are negligible when the length of the pillar is much greater than its radius. At the same time, the bridging of low-resistance magnetic material can prevent poling the piezoelectric and upset the open circuit condition. It is reasonable to use most high-resistive magnetostrictive materials such as appropriate ferrites.

Our studies indicate that flexural deformation gives rise to a decrease in the strength of ME interaction for a free-standing bilayer. The decrease is maximum for YIG volume fraction approximately equal to 0.5. The substrate thickness dependence of FMR line shift reveals a maximum for the substrate thickness that is approximately equal to that of the bilayer. Further increase in the substrate thickness leads to a decrease in ME effect due to clamping effect produced by the substrate. However, rate of decrease in ME coupling strength is considerably lower than that for the case when neglecting the flexural strains. The results are of interest for novel ferrite–piezoelectric microwave devices.

Acknowledgments

This work was supported by the Russian Foundation for Basic Research and Programs of Russian Ministry of Education and Science.

References

1. N.A. Pertsev, A.G. Zembilgotov, A.K. Tagantsev, "Effect of mechanical boundary conditions on phase diagrams of epitaxial ferroelectric thin films," Phys. Rev. Lett., **80**, 1988 (1998).

2.   C. Deng, Y. Zhang, J. Ma, Y. Lin, C.-W. Nan, "Magnetoelectric effect in multiferroic heteroepitaxial $BaTiO_3$–$NiFe_2O_4$ composite thin films," Acta Materialia, **56**, 405 (2008).

3.   V.M. Petrov, G. Srinivasan, M.I. Bichurin, T.A. Galkina, "Theory of magnetoelectric effect for bending modes in magnetostrictive-piezoelectric bilayers," J. Appl. Phys. **105**, 063911 (2009).

4.   H. Zheng, J. Wang, S.E. Lofland, Z. Ma, L. Mohaddes-Ardabili, T. Zhao, L. Salamanca-Riba, S.R. Shinde, S.B. Ogale, F. Bai, D. Viehland, Y. Jia, D.G. Schlom, M. Wuttig, A. Roytburd, R. Ramesh, "Multiferroic $BaTiO_3$-$CoFe_2O_4$ nanostructures," Science, **303**, 661 (2004).

5.   G. Liu, C.-W. Nan, J. Sun, "Coupling interaction in nanostructured piezoelectric/magnetostrictive multiferroic complex films," Acta Mater. **54**, 917 (2006).

6.   S. Shastry, G. Srinivasan, M.I. Bichurin, V.M. Petrov, A.S. Tatarenko, "Microwave magnetoelectric effects in single crystal bilayers of yttrium iron garnet and lead magnesium niobate-lead titanate," Phys. Rev., **B70**, 064416 (2004).

7.   C.-W. Nan, G. Liu, Y.H. Lin, H. Chen, "Magnetic-field-induced electric polarization in multiferroic nanostructures," Phys. Rev. Lett., **94**, 197203 (2005).

8.   V.M. Petrov, G. Srinivasan, M.I. Bichurin, A. Gupta, "Theory of magnetoelectric effects in ferrite piezoelectric nanocomposites," Phys. Rev., **B75**, 224407 (2007).

9.   M.I. Bichurin, V.M. Petrov, G. Srinivasan, "Low-frequency magnetoelectric effects in ferrite-piezoelectric nanostructures," JMMM, **321**, 846 (2009).

10. M.I. Bichurin, V. M. Petrov, T. A. Galkina, "Microwave magnetoelectric effects in bilayer of ferrite and piezoelectric," Eur. Phys. J. Appl. Phys., **45**, 30801 (2009).

Index

Color Insert

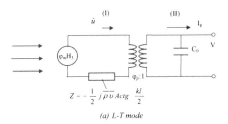

(a) L-T mode

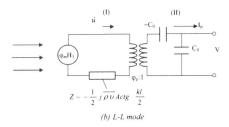

(b) L-L mode

Figure 5.2. Magneto-elastic-electric bi-effect equivalent circuits for (a) L–T mode, where $\varphi_{\mathrm{m}} = 2A_2\dfrac{d_{33,m}}{s_{33}^H}$, $\varphi_{\mathrm{p}} = \dfrac{wd_{31,p}}{s_{11}^E}$; $C_0 = \dfrac{lw}{t_1}\varepsilon_{33}^T$; and (b) L–L mode, where $\varphi_{\mathrm{m}} = 2A_2\dfrac{d_{33,m}}{s_{33}^H}$; $\varphi_{\mathrm{p}} = \dfrac{A_1 g_{33p}}{ls_{33}^D\overline{\beta}_{33}}$; $C_0 = \dfrac{A_1}{l\overline{\beta}_{33}}$.

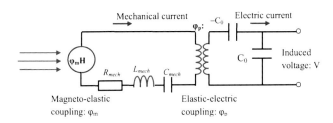

Figure 5.3 Magneto-elastic-electric equivalent circuits for L–L at resonance [14,18], where $L_{mech} = \dfrac{\pi Z_0}{8\omega_s}$, $C_{mech} = \dfrac{1}{\omega_s^2 L_{mech}}$, and $Z_0 = \overline{\rho\upsilon}A_{lam}$, $R = \dfrac{\pi Z_0}{8Q_m}$.

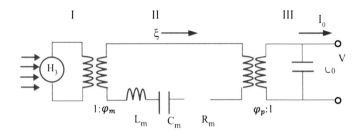

Figure 5.6 Equivalent circuit of ME C–C mode.

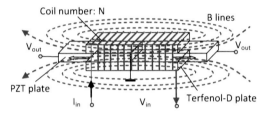

(a) Configuration of ME transformer.

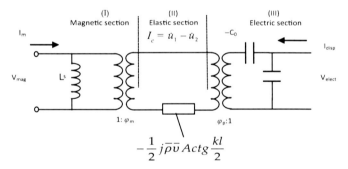

(b) Equivalent circuit at low frequency.

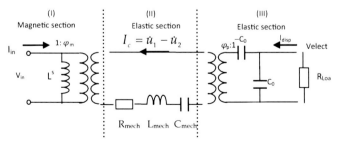

(c) Equivalent circuit at resonance.

Figure 5.7 ME transformer.

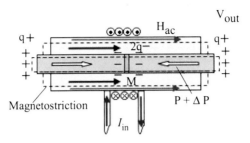

(a) ME gyration configuration.

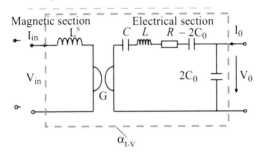

(b) Equivalent circuit of ME gyrator equivalent circuit modelof ME gyrator under resonance drive and free-condition, where $G = \varphi_m / \varphi_p$;

$$\varphi_m = \frac{NA_m d_{33,m}}{s_{33}^H l}; \quad \varphi_p = \frac{2A_p g_{33,p}}{l s_{33}^D \bar{\beta}_{33}}; \quad L^s = \frac{\mu^s N^2 A_m}{l}; \quad R = \frac{\pi Z_0}{8 Q_m \varphi_p^2}; \quad L = \frac{\pi Z_0}{8 \omega_s \varphi_p^2};$$

$$C = \frac{\varphi_p^2}{\omega_s^2 L_{mech}}; \quad C_0 = \frac{2A_p}{l \bar{\beta}_{33}}; \quad Z_0 = \bar{\rho}\bar{v} A_{lam}; \quad \text{and} \quad \omega_s = \frac{\pi \bar{v}}{l}. \quad \text{The parameters}$$

A_m, A_p, and A_{lam} are the cross-sectional areas of the magnetostrictive layers, piezoelectric layer, and laminate, respectively; l is the length of the laminate; $\bar{\rho}$ and $\bar{v}$ are the mean density and acoustic velocity of the laminate; μ^s is the magnetic permeability of the magnetostrictive layer under constant stress; Q_m is the mechanical quality factor of the laminate; and N is the turn number of coils.

Figure 5.8. ME gyrator [47].

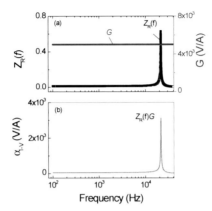

Figure 5.9 Calculations using Eq. 5.24 for (a) ideal gyration coefficient G, and frequency transfer function Z_R (f); and (b) I–V conversion coefficient a_{I-V}, as a function of drive frequency f.

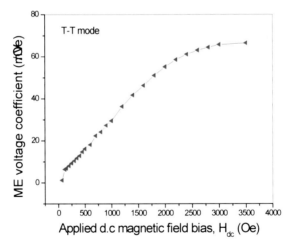

Figure 5.14 ME voltage coefficient as a function of H_{dc} [24].

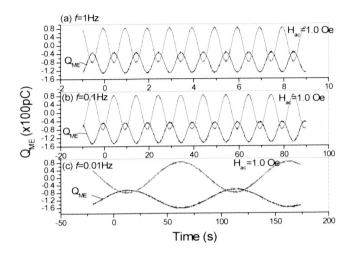

Figure 5.18 Extremely low-frequency responses of bimorph-type ME sensor.

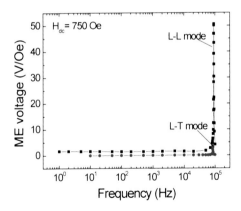

Figure 5.19 ME voltage coefficients of Fe-Ga/PMN–PT laminates as a function of magnetic field frequency [39].

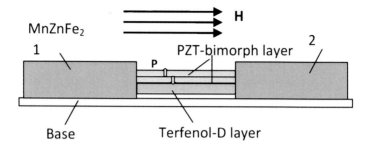

Figure 5.20 Configuration of three-phase $MnZnFe_2O_4$/Terfenol-D/PZT ME laminate [42].

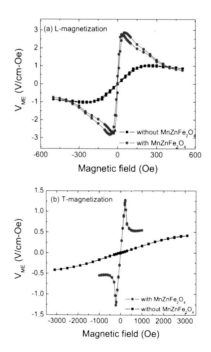

Figure 5.21 High-μ phase's effect on ME voltage coefficients in three-phase ME composites.

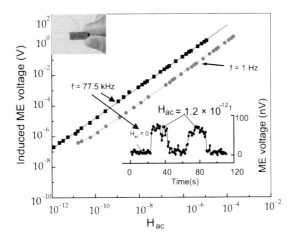

Figure 5.22 Limit magnetic field sensitivity of Terfenol-D/PMN–PT ME laminate [22].

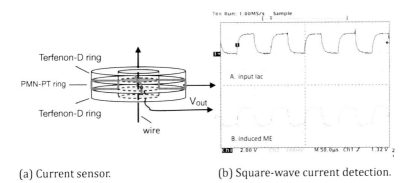

(a) Current sensor.   (b) Square-wave current detection.

Figure 5.25 ME current sensor [37].

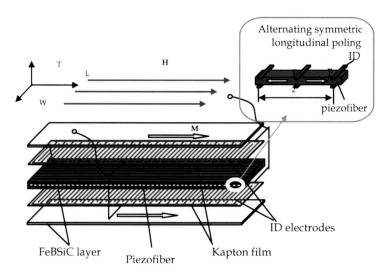

Figure 5.28 (2–1) connectivity Metglas/PZT fiber laminate [40].

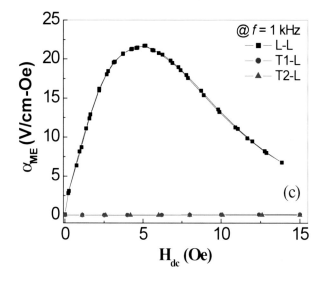

Figure 5.29 ME field coefficient as a function of dc magnetic bias for longitudinal (L), width ($T1$), and thickness ($T2$) magnetization [40].

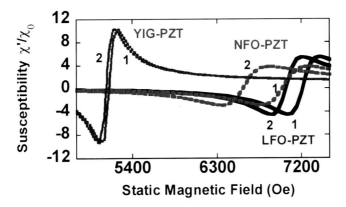

Figure 6.13 Theoretical FMR profiles in ferrite–PZT bilayers at 9.3 GHz for $E = 0$ (labeled — 1) and $E = 300$ kV/cm(−2). E and H are perpendicular to the bilayer [22].

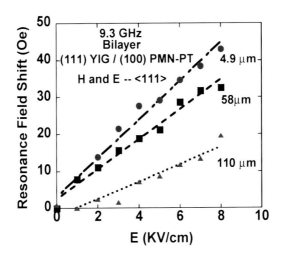

Figure 6.14 Data on FMR field shift vs. E for a YIG/PMN–PT bilayer.

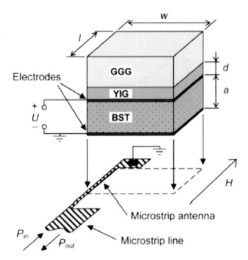

Figure 6.15 Diagram showing the schematics of a YIG–BST layered system for hybrid wave generation [50].

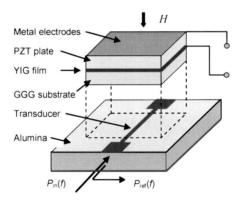

Figure 6.17 YIG–PZT resonator [47].

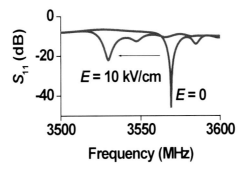

Figure 6.18 Electric field tunable 13 μm YIG–PMN–PT resonator with perpendicular magnetization (H = 3010 Oe) [47].

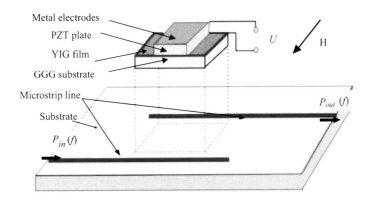

Figure 6.19 A magnetoelectric (ME) band-pass filter. The ME resonator consisted of a 110 μm thick (111) YIG on GGG bonded to PZT [52].

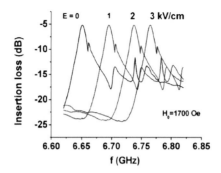

Figure 6.20 Loss vs. *f* characteristics for a series of *E* for the YIG–PZT filter [52].

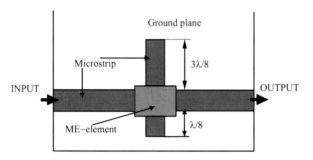

Figure 6.21 Diagram showing the schematics of a YIG/PZT phase shifter [51].

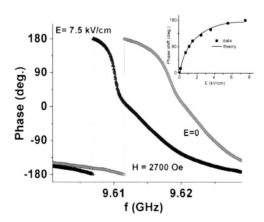

Figure 6.22 The phase angle ϕ vs. frequency *f* at 9.6 GHz for *E* = 0 and 7.5 kV/cm [51].

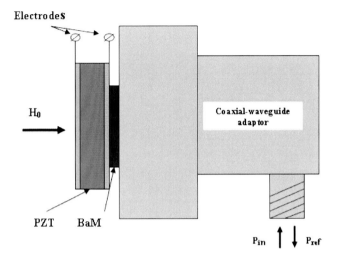

Figure 6.23 Measurement cell for mm-wave ME effects in BaM–PZT bilayers [53].

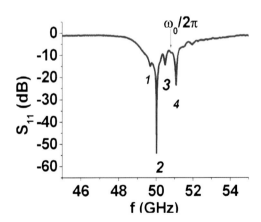

Figure 6.24 Magnetostatic forward volume modes in BaM [53].

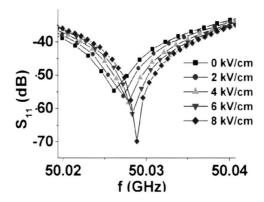

Figure 6.25 Frequency shift vs. *E* for mode 2 in Fig. 6.20 for BaM–PZT bilayer [55].

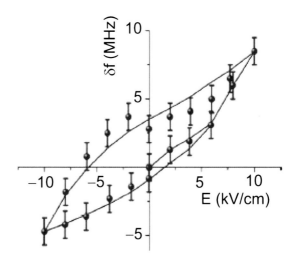

Figure 6.26 Shift vs. *E* for 95 μm thick BaM and PZT bilayer [55].

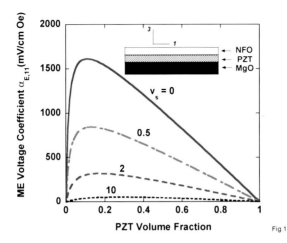

Figure 7.1 Diagram showing a nickel ferrite (NFO)–lead zirconate titanate (PZT) nanobilayer in the (1,2) plane on MgO substrate. It is assumed that PZT is poled with an electric field E_1 along 1, the bias magnetic field H_1 and ac magnetic field δH_1 are along axis-1, and the ac electric field δE_1 is measured along direction-1. Estimated PZT volume fraction dependence of in-plane longitudinal ME voltage coefficient $\alpha_{E,11} = \delta E_1/\delta H_1$ is shown for a series of volume fraction v_s for MgO.

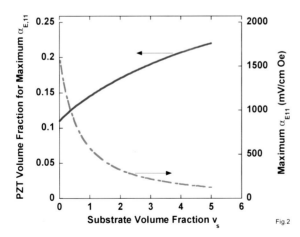

Figure 7.2 Variation with substrate volume fraction v_s of the peak value of ME voltage coefficient $\alpha_{E,11}$ in Fig. 7.3 and the corresponding PZT volume fraction.

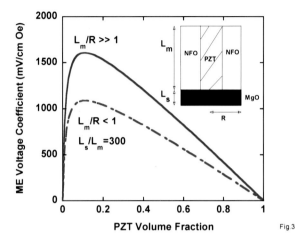

Figure 7.5 A nanopillar of NFO and PZT of radius R and length L_m on an MgO substrate of thickness L_s. All the electric and magnetic fields are assumed to be along the axis (direction-3) of the pillars. Estimated PZT volume fraction dependences of ME voltage coefficient $\alpha_{E,33}$ for NFO–PZT nanopillar are shown.

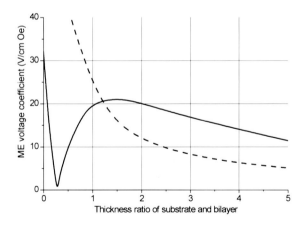

Figure 7.9 Dependence of peak ME voltage coefficient for the bilayer of NFO and PZT at bending mode on ratio of SrTiO$_3$ substrate thickness to bilayer thickness are presented for PZT volume fraction 0.6. For comparison, the ME voltage coefficient at longitudinal mode is shown (dash line).